Lecture Notes in Economics and Mathematical Systems

644

Founding Editors:

M. Beckmann
H.P. Künzi

Managing Editors:

Prof. Dr. G. Fandel
Fachbereich Wirtschaftswissenschaften
Fernuniversität Hagen
Feithstr. 140/AVZ II, 58084 Hagen, Germany

Prof. Dr. W. Trockel
Institut für Mathematische Wirtschaftsforschung (IMW)
Universität Bielefeld
Universitätsstr. 25, 33615 Bielefeld, Germany

Editorial Board:

H. Dawid, D. Dimitrow, A. Gerber, C-J. Haake, C. Hofmann, T. Pfeiffer,
R. Slowiński, W.H.M. Zijm

Julia Drechsel

Cooperative Lot Sizing
Games in Supply Chains

 Springer

Julia Drechsel
Germany
julia.drechsel@uni-due.de

Zugl.: Duisburg-Essen, Univ., Diss., 2009

ISSN 0075-8442
ISBN 978-3-642-13724-2   e-ISBN 978-3-642-13725-9
DOI 10.1007/978-3-642-13725-9
Springer Heidelberg Dordrecht London New York

Library of Congress Control Number: 2010934000

Cover design: WMXDesign GmbH, Heidelberg, Germany

Printed on acid-free paper

Springer is part of Springer Science+Business Media (www.springer.com)

Preface

This book results from my work as a research assistant at the Universities of Freiberg and Duisburg-Essen. Starting this research was very challenging due to the highly mathematically coined literature regarding game theoretical topics. The literature shows that cooperative game theory is well studied from a theoretical point of view, but that fields of application are fairly limited up to now. Hence, the following research question arose: What can be done to make concepts from cooperative game theory more applicable to practical problems?

The field of supply chain management is predestined for many kinds of cooperation because companies will be more successful if they consider relationships to suppliers, customers, and/or competitors. Therefore, this work focuses on cooperations in supply networks (horizontal and vertical cooperation) concerning joint ordering and/or joint production.

Problems of ordering and production for single decision makers are well studied in operations research and are used as a basis to develop cooperative models that display situations of cooperative decision making. Being one of the most essential problems in cooperations, the allocation of joint profits or costs is the fundamental question in cooperative game theory and will be the basic question answered by this work as well.

In both places of my doctorial studies, Freiberg and Duisburg, I had valuable support without which this work would not have been possible. First of all, I thank Alf Kimms for providing me the chance to work on this topic, his guidance, challenging discussions, and the possibility to continue my work in Duisburg.

I express my gratitude to Sarah Bretschneider, Demet Çetiner, Michaela Graf, Klaus-Christian Maassen, and Michael Müller-Bungart for providing me such a great research and working atmosphere in Freiberg and Duisburg – all of you helped me through struggles with research, programming, or teaching, as well as sometimes non-work related things, which was always valuable and helpful and made these four years a wonderful time. Furthermore, I would like to thank Annette Hoffmann and Nadine Krumpholz for welcoming me in Duisburg, Stefanie Kockerols for providing me help with the literature research and Çağdaş Özgür for a constant LATEX support.

I am deeply indebted to Ingmar Schaaff for his love, his patience, and for keeping me grounded whenever it was needed. Finally, I thank my parents, Nicole and

Stephan Drechsel, for making so many things possible, for their support, and their trust.

Duisburg   *Julia Drechsel*
May 2010

Contents

List of Figures

List of Tables

Chapter 1
Introduction

Cooperations are gaining more and more importance in the field of supply chain management due to highly complex supply network structures. Therefore, it is not sufficient to plan operations in terms of isolated decision making. The benefit of cooperation is extensively discussed in the literature. While examining cooperations, many decision problems arise; e.g., choosing the right partners, evaluating the success of a cooperation, allocating costs and profits among the partners. The last topic, in particular, is of great importance because a right allocation assures stability and fairness in a cooperation.

This question seems to be some kind of accounting problem, but, however, cooperative game theory can give an answer to it as well. Methods from cooperative game theory are particularly useful if prices are not determined externally by the market, but are set internally by mutual agreement or administrative decision (see Young 1994, p. 1194). The core is one of the most popular concepts for cost/profit allocation. It identifies shares in a way that each partner in the cooperation does not have any incentive to leave the cooperation and work on its own or form a smaller coalition due to higher payoffs or lower costs, respectively. Hence, the core is an instrument to guarantee the stability of a cooperation.

The theoretical foundation of cooperative game theory is well established. A great interest in applications for these research results emerges. Due to the mathematical background of cooperative game theory, the literature is concentrating on theoretically proofing special properties for the considered games (e.g., concavity, subadditivity, or emptiness of the core). Some of these findings are important to know but of lower relevance for practical applications; e.g., it is nice to know that the core for an inventory game is not empty but it would be much more relevant to know how to calculate a concrete core allocation.

In this work, we start focusing on problems from inventory management where several companies or business units are facing the same ordering decision (order the same product or raw material from the same supplier under the same conditions). Such problems are well studied in the literature, though mostly from the point of view of a single decision maker. With the goal to use economies of scale, however, they might cooperate and place orders together. In practice, this topic is not only relevant for private companies, but also for public bodies. They are forming so-called purchasing alliances to bundle resources, use quantity discounts, and save

J. Drechsel, *Cooperative Lot Sizing Games in Supply Chains*, Lecture Notes in Economics and Mathematical Systems 644, DOI 10.1007/978-3-642-13725-9_1, © Springer-Verlag Berlin Heidelberg 2010

costs. First approaches for the classical lot sizing problem to find allocations in the core exist in the literature. This problem, in particular, is relatively easy to handle because it can be modeled with a closed formula. For the dynamic lot sizing problem, a very special algorithm exists in the literature which can only be used for this kind of problem.

Therefore, one aim of this thesis is to develop a general algorithm for computing core allocations. Core allocations are defined by a bunch of restrictions. Hence, it is possible to formulate a constraint satisfaction problem to compute allocations in the core. The number of these constraints is exponentially rising with a growing number of coalition members. To reduce this big model and speed up computation time, we propose a row generation procedure. The idea of such an approach is to solve a relaxed constraint satisfaction problem (master problem) by neglecting most of the restrictions first. In a second step, we solve a subproblem to find a restriction (that we left out in the first step) for which the optimal solution from step one is not feasible. According to this, we add the identified restriction to the master problem and solve it again. The algorithm stops if there is no restriction found in the subproblem violating the last solution from the master problem. First computational tests show that the algorithm delivers promising results. An extensive computational study should show the efficiency of the algorithm.

Furthermore, it needs to be analyzed whether it is possible to use this algorithm also for other (more complex) cooperative optimization problems. Possible extensions that will be considered include:

- Economic lot sizing problems with uncertain demand
- Capacitated lot sizing problems
- Multilevel lot sizing problems

While exploring advantages and disadvantages of the core as the right allocation method, another field of investigation comes up: Can the basic procedure of the algorithm also be adopted to other allocation schemes from cooperative game theory; e.g., core variants like the ϵ-core or the least core? In addition to the property of stability, we will discuss ideas to compute fair allocations as well.

The major contribution of this thesis is to investigate cooperative situations in supply chain management and show their practical value. By developing appropriate models, we can check theoretical properties of these games. In the next step, a general algorithm for computing core allocations will be introduced to make concepts of cooperative game theory, especially the core, applicable for practical use. This algorithm will be treated under the special aspects mentioned above.

The course of this book can be illustrated by Fig. 1.1. Chapters 2–4 mark the two directions the work is based on: Chap. 2 will present selected topics of cooperative game theory that are needed to understand the explanations and discussions following in the later chapters. An important contribution is covered by Sect. 2.3 where, after presenting the concept of the classical core, several promising variants will be introduced. In Chap. 3, a general algorithm will be developed to compute core cost allocations. After presenting the algorithmic theory, Chap. 4 will describe decision problems in supply chain management where cooperations play an important role

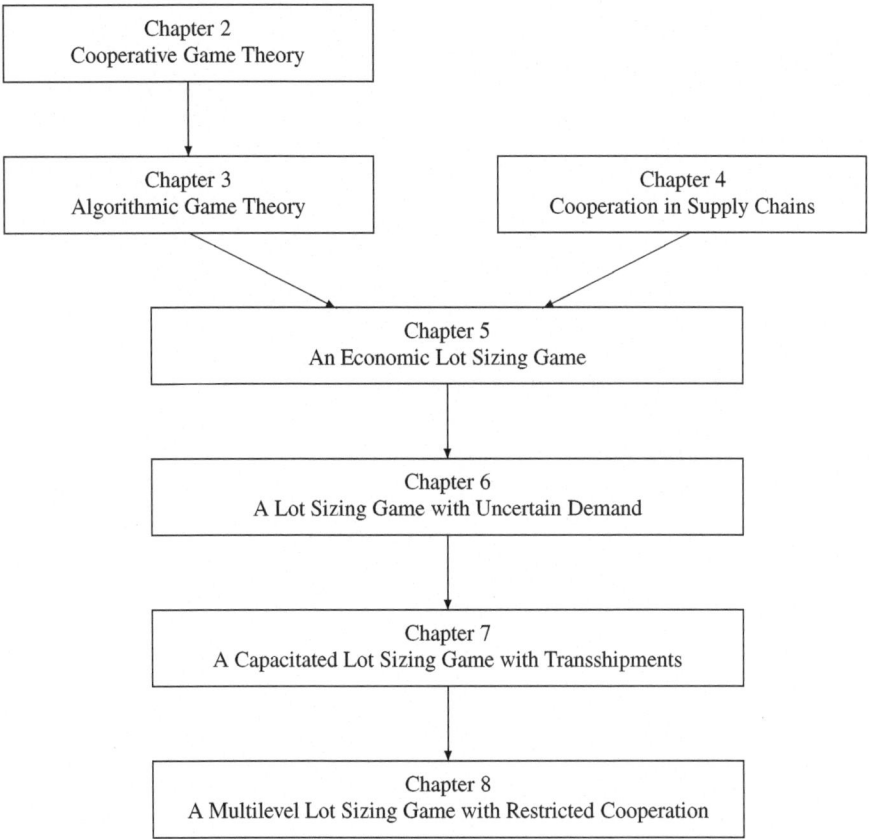

Fig. 1.1 Structure of the thesis

and, hence, cooperative game theory would provide interesting solution approaches to handle such problems. Chapters 5–8 combine the two streams and deal with applications, especially from the field of supply chain management, for the developed core concepts and the algorithm. We will present a very simple form of a horizontal cooperation in Chap. 5 and will discuss in detail more complex situations with either uncertain demand (Chap. 6), cooperative production and transshipments (Chap. 7), or multilevel supply situations (Chap. 8).

Chapter 2
Selected Topics in Cooperative Game Theory

This chapter shall provide the theoretical basis coming from cooperative game theory. We start with a survey of historical developments in the field of game theory. After this short excursion, we introduce cooperative games and their properties. As the fundamental contribution of cooperative game theory are methods to allocate cooperative profits or costs, we present the most prominent allocation methods. Thereby, we concentrate on the core and its variants as this method will be applied in the following chapters.

2.1 History of Game Theory

Let us start with answering two questions: What is game theory and where does it come from? One general definition is provided by Myerson (1991): "*Game theory can be defined as the study of mathematical models of conflict and cooperation between intelligent rational decision-makers.*" The essential point is that individual decisions will influence each other's welfare (which may stand for profits or costs). Binmore (1992) describes that a *game* is being played whenever people interact with each other. If you and your supervisor negotiate about your last year's bonus, it is called a game in the sense of game theory. When several small companies build up an alliance or cooperation to gain more power in handling customers or suppliers, it is a game. It becomes clear that the notion "game" is somewhat misleading because it stands for an event where people interact. Hence, Myerson (1991) suggests to use terms like "conflict analysis" or "interactive decision theory" for this stream of research. However, the term "game theory" is most common in modern literature.

Modern game theory is mostly based on the seminal book of John von Neumann and Oskar Morgenstern published in 1944 (in 2004 the fourth edition was published, see von Neumann and Morgenstern 2004). There are earlier works by Edgeworth (1881), von Neumann (1928), Cournot (1838), Bertrand (1883), and Zermelo (1913) but they are lacking a uniform theory regarding strategic behavior of interactive decisions. Thus, the book of von Neumann and Morgenstern paved the way for an extensive research in the field of game theory. First expectations of revolutionizing social sciences with insights of game theory were detached by important

J. Drechsel, *Cooperative Lot Sizing Games in Supply Chains*, Lecture Notes
in Economics and Mathematical Systems 644, DOI 10.1007/978-3-642-13725-9_2,
© Springer-Verlag Berlin Heidelberg 2010

applications, especially in the field of economics. Von Neumann and Morgenstern (1944) provide an abstract mathematical formulation for conflict situations. They build up an extensive theory coining notions as the normal or extensive form of games, concepts of pure and mixed strategies, coalitional games, stable sets, etc. Apart from that, they develop the utility theory.

Von Neumann and Morgenstern distinguish between two approaches to the theory of games: The first is called *strategic* or *non-cooperative* game theory, whereas the second stream is called *coalitional* or *cooperative* game theory. A non-cooperative game specifies all possible actions for every involved decision maker. Based on this knowledge, we search for a strategy that is best for each decision maker. Recall, it is assumed that the results for every party depend on the strategy of the other parties. In the case of cooperative games, von Neumann and Morgenstern deal with patterns of coalition formation under rational behavior in particular. Still today, the field of game theory is divided into those two streams of cooperative and non-cooperative game theory.

Following the work of von Neumann and Morgenstern, various contributions to the topic of game theory were published since then. Therefore, we present only selected, very important contributions in this section without claiming completeness. A scientist contributing substantially to both streams of game theory is John Nash. In Nash (1950, 1951), he succeeds in developing a general formulation of the equilibrium idea which leads to the well known notion of the *Nash equilibrium* used in non-cooperative games. The Nash equilibrium is a combination of strategies where no player could increase its expected payoff by unilaterally deviating from its strategy. In Nash (1953), he also makes a contribution to cooperative game theory. He suggests two different methods to derive solutions for a two-person cooperative game: In the first one, the cooperative game is reduced to a non-cooperative game while making the players' steps of negotiation moves in the non-cooperative model. The second approach is based on axioms that describe desirable properties and lead to a uniquely solution.

In the 1950s, the famous *prisoner's dilemma* was introduced as a special example for non-zero-sum games. Luce and Raiffa (1957) is one of the first references that discusses the prisoner's dilemma extensively.

In the stream of cooperative game theory, Gillies (1959) and Shapley (1953), provide an important basis. Gillies (1959) invents the *core* as a general solution method for cooperative n-person games. At the same time, Shapley (1953) introduces the *Shapley value* as a further solution concept. While the core in most of the cases represents a set of possible allocations with specific properties, the Shapley value provides a unique allocation following other criteria. We will study those two solution concepts and some further ones in Sect. 2.3 in detail.

In this early time of game theoretical research, most applications were found in economics, politics, military, and first steps in evolutionary biology (see Lewontin 1961). In the area of economic applications, Shubik (1962) suggests to use methods from cooperative game theory to solve cost accounting problems in a company. Following the time line, Aumann and Maschler (1964) develop a further allocation method which they call the *bargaining set*. The trend to develop more and more different solution concepts for cooperative game theory that are based on varying properties has continued until now. None of the solution concepts appeared to be superior because all of them have their advantages and disadvantages. In contrast, the analysis of non-cooperative games is mostly based on the Nash equilibrium. However, research continues, for instance, in refining the concept of the Nash equilibrium: Selten (1965), in particular, contributes to this while introducing the notion of the *subgame perfect equilibrium*.

Most of the research today is following the classification of Harsanyi (1966) regarding *cooperative* vs. *non-cooperative* game theory. He defines a cooperative game as a game where commitments are fully binding and enforceable. Commitments can be, for instance, agreements, promises, or threats. A non-cooperative game has no binding commitments. Harsanyi (1966) states that any cooperative game can be replaced by a non-cooperative one. To execute this, we have to incorporate the commitments into the available strategies and the penalties into the payoff matrix when commitments are violated. However, Harsanyi (1966) does not suggest to deal only with non-cooperative games because this transformation greatly increases the size of the payoff matrix. In Harsanyi and Selten (1988), this definition is refined: The underlying assumption for a non-cooperative game is that the players are unable to make enforceable agreements except in the case where the extensive form of the game explicitly gives them an ability to model such commitments. According to that, a cooperative game contains the assumption that the players are able to make binding agreements, even if this possibility is not explicitly shown in the extensive form of the game (see Harsanyi and Selten 1988, p. 3). Such binding agreements are prevalent in many areas of economics. Almost every seller-buyer relation is characterized by some binding principles fixed in a contract – this does not only hold for one stage relations but also for multi-stage transactions. Most of these contracts include monetary penalties that discourage the partners from breaking the agreements, thus making the contract binding.

Apart from this, Harsanyi makes substantial contributions to the theory of games with incomplete information (see Harsanyi 1967, 1968a,b). A further refinement of the Nash equilibrium is developed by Selten (1975) under the name *trembling hand perfect equilibrium*. The joint work Harsanyi and Selten (1988) continues the research regarding the refinement of the Nash equilibrium and was crowned with the Nobel prize in economic sciences for Harsanyi, Nash, and Selten in 1994.

We already specified the term *game*. Let us examine some more notions that appear very often when dealing with aspects of game theory. The number of involved decision makers in a game is not bounded and we call the decision makers *players*. Myerson (1991) formulates two important assumptions regarding the properties of the players:

- Rationality: Each player's objective is to maximize its expected payoff (measured on some utility scale). Hence, there must be a possibility to assign utility numbers to all possible outcomes of the game (see the expected-utility maximization theorem by von Neumann and Morgenstern 2004, p. 30). This means the player's only interest is the maximization of its own payoff, no matter the other players' payoffs.

- Intelligence: It is assumed that a player knows everything about the game. This means the players can make the same inferences as ourselves. The same holds for cases where a theory is applied to a game. In this case, every player is aware of this knowledge.

Today, game theory is seen as an independent area of decision theory. To be more specific, game theory is an extension of classical decision theory to the case of more than one decision maker. Morgenstern (1963, p. 77), distinguishes between the case in classical decision theory, where the outcome of decisions of one player depends on (random) external events (not manipulable by the player) and the case of game theory, where two players interact; i.e., outcomes depend on the opponents and maybe additionally on (random) external events.

Until the 1970s, game theory was mostly a mathematical discipline. Later on, game theory was also applied to economical problems, like for instance antitrust analysis, monetary policy, and business problems. Any form of competition or cooperation in a company or between companies provides decision problems that can be analyzed with the help of game theory methods. Players might be companies, business units, suppliers, customers, or even employees. A groundbreaking work is done in the book of Brandenburger and Nalebuff (1996) because they use game theoretical models in the background while approaching them through real-life stories. By using game theory to explain success or failure of decisions, they summarize the lessons in checklists to make the methods applicable for non-scientists, too. Chatterjee and Samuelson (2001) and Jost (2001) provide each an extensive survey of game theoretical contributions in the field of business applications and present discussions about limits of application as well.

We do not claim to give a complete survey with the above paragraphs. To gain extensive and superior knowledge in the field of game theory, we would like to refer exemplary to the following textbooks: Myerson (1991), Fudenberg and Tirole (1991), Binmore (1992), Owen (2001), and Rasmusen (2007).

2.2 Basics in Cooperative Game Theory

This section shall give an introduction into cooperative game theory. We will learn about the definition and properties of cooperative games as a basis to analyze allocation methods in the following sections. Following Myerson (1991, p. 370), we will use the term *cooperation* in the meaning of 'players interacting with a common purpose'. The main subjects of cooperative game theory are to analyze whether there are incentives to cooperate and to allocate costs (or profits) accrued in the cooperation among the participating players.

The contributions of John Nash coin the beginning of research in the field of cooperative game theory: Nash (1951)'s work is based on von Neumann and Morgenstern (1944)'s cooperative n-person game. Nash (1951) argues that the process of bargaining among players may result in cooperative actions. Hence, he suggests using the Nash equilibrium to analyze non-cooperative as well as cooperative games. In the bargaining process, each player follows its individual bargaining strategy that maximizes its utility (see Myerson 1991, especially Chap. 8, for details about such a transformation). However, a consequent transformation might lead to a large set of equilibria. Apart from this transformation, Nash (1950, 1953) suggest a solution concept called the Nash bargaining solution. It is based on the assumption that the profit allocation depends only on the profits the players would expect if negotiation fails and on the set of profit allocations that are jointly feasible. It is possible to extend this approach to an n-person Nash bargaining solution but its application is questionable and unusual because the concept completely ignores the possibility of cooperation among subsets of players and their (negotiation) power. For Myerson (1991), the main assumption for differentiating between cooperative and non-cooperative game theory is that players can negotiate effectively in cooperative games. *Effective negotiation* means that in case of feasible strategy changes that would benefit all coalition members, they would agree to make this change as long as it does not contradict agreements made with players outside the coalition (see Myerson 1991, Sect. 9.1, for more details).

For substantial introductions into cooperative game theory, we refer exemplary to the following textbooks: Myerson (1991), Osborne and Rubinstein (1994), Curiel (1997), Owen (2001), and Peleg and Sudhölter (2007).

2.2.1 A Cooperative Game

A fundamental assumption in cooperative game theory is whether or not the result of a cooperation can be quantified and transferred (without gain or loss) among the players. Therefore, the assumption of *transferable utility* (TU) is often used in cooperative game theory to handle interactions in-between different coalitions. With transferable utility, we mean a commodity (like money) that can be freely transferred among the players without any third instance. While transferring, any player's utility cost/profit increases one unit for every unit of the commodity it gets (see e.g., Myerson 1991, Sects. 8.4 and 9.2). The transferable utility helps to define the characteristic function of a cooperative game. In the following course of this book, we will solely discuss games with transferable utility. For games with non-transferable utility (NTU), compare, for instance, Myerson (1991) and Peleg and Sudhölter (2007). According to the use of transferable utility, we assume that the players prefer less money to more money in the case of costs (vice versa in the case of profits) and that the players are risk averse in their preferences for money (see Maschler 1992, p. 594). Hence, it is possible to neglect individual utility functions. Literature regarding cooperative game theory mostly uses payoffs that represent the utility functions.

A *cooperative game* with transferable utilities (short: TU game) is characterized by two main factors: $N = \{1, 2, \ldots, |N|\}$ is the given set of players and c is the characteristic function. Players are decision makers and we call every subset $S \subseteq N$ of cooperating players a *coalition*. If two players are not cooperating, they belong to different coalitions; i.e., there is only one occurrence of cooperation. N is called the grand coalition. As the definition of game theory in Sect. 2.1 implies, a game emerges when two or more players interact. Analyzing possible strategies and solutions of two-player-games is manageable, but dealing with more than two decision makers increases the complexity of the problem. Obviously, this is due to the exponentially rising number of coalitions $(2^{|N|})$ that has to be taken into account when the number of players increases. For the further discussion, we assume that the set of players is finite. Games with an infinite set of players are called *nonatomic games* (see e.g., Owen 2001).

The second factor, the *characteristic function*

$$c : 2^N \to \mathbb{R}$$

assigns a cost or profit value (representing the total amount of transferable utility) to each coalition $S \subseteq N$ which determines the best outcome for the coalition S if the players in S cooperate without the players in $N \setminus S$. In general, it is assumed that for an empty set

$$c(\emptyset) = 0.$$

Assume that the players form the grand coalition and want to divide total costs $c(N)$ among themselves after some kind of bargaining process. The outcome will heavily depend on the power structure in the grand coalition. A player's power could be interpreted as its ability to help or hurt any (sub-)set of players by agreeing or refusing to cooperate (see Myerson 1991, p. 427). Hence, a characteristic function can be seen as a summary description of the power structure of the game.

In short, a game is defined by the pair

$$\Gamma := (N, c).$$

The terms *game in coalitional form* or *coalitional game* can be used instead of the notion characteristic function. It is sufficient to give the characteristic function to define a cooperative game because the player set is given implicitly via the dimension of the characteristic function. For special representations of the presented coalitional form based on offensive and defensive threats among the coalitions, see Myerson (1991, p. 423).

The characteristic function can be interpreted as profits or costs – we will speak of profit and cost games, respectively. In a profit game, players prefer a higher outcome for themselves, whereas in a cost game, they favor lower amounts. The symbol v is usually used for profit functions and c for cost functions. Most of the time, profit and cost games can be handled the same way. An exception to this is the existence of restricted cooperation (see Chap. 8 for a further discussion on restricted cooperation). For a mathematical comparison and discussion of profit and cost games, see also Bilbao and Martínez-Legaz (2002). It is also customary in game theory and its applications to convert a cost game into a cost savings game with

$$v(S) = \sum_{i \in S} c(\{i\}) - c(S) \quad \text{for all } S \subseteq N$$

(see Young et al. 1982, p. 464, or Lemaire 1984, p. 68) or to make the transformation with the help of the so-called *dual game* ($c(S)$ is the dual game of $v(S)$)

$$c(S) = v(N) - v(N \setminus S) \quad \text{for all } S \subseteq N$$

(see Curiel 1997, p. 16, for a numerical example). Obviously, $v(\emptyset) = c(\emptyset) = 0$ holds for both transformations.

In the course of this thesis, we will deal with cost games (N, c), unless stated otherwise, and will not use one of these transformations. Thus, positive values of the characteristic function represent costs and negative values could be interpreted as profits. Obviously, it is possible to use other quantifiable units than costs or profits, too.

π_i denotes the cost (or profit) share that will be allocated to player i. These shares should be computed in such a way that the vector $\pi = (\pi_1, \pi_2, \ldots, \pi_{|N|})$ gives an allocation of the total costs for the grand coalition $c(N)$. This vector is an element of the $|N|$-dimensional linear space R^N.

2.2.2 Properties of Cooperative Games

Some important properties to classify characteristic functions are introduced in this section. Those properties are used to classify cooperative games and to derive insights about the application of solution concepts.

A cooperative game is *monotone* if and only if the cost function does not decrease with the number of players in the coalition:

$$c(S_1) \leq c(S_2) \qquad S_1 \subseteq S_2 \subseteq N. \tag{2.1}$$

The same interpretation can be used to describe monotone profit games:

$$v(S_1) \leq v(S_2) \qquad S_1 \subseteq S_2 \subseteq N.$$

If an opportunity to cooperate among a set N of players exists, it is interesting to find out how the "optimal" coalition structure looks like. Structure in this sense stands for a *partition*

$$S_1, \ldots, S_m \quad \text{with} \quad \bigcup_h S_h = N \quad \text{and} \quad S_h \cap S_{h'} = \emptyset \quad \text{for all } h \neq h'.$$

Let $\mathcal{P}(N)$ denote the set of all partitions of N. An "optimal" partition means a partition of players with minimum total costs ($\min \sum_h c(S_h)$). The size m of the partition is not known in advance. If we can prove, however, that the characteristic function is subadditive, then the grand coalition is the optimal partition ($m = 1$ and $S_1 = N$). Thus, *subadditivity* indicates whether two players or two disjoint coalitions have an incentive to cooperate because they have lower costs while cooperating compared to individual activity:

$$c(S_1) + c(S_2) \geq c(S_1 \cup S_2) \qquad S_1, \ S_2 \subseteq N, \ S_1 \cap S_2 = \emptyset. \qquad (2.2)$$

Maschler (1992) argues that in case of a bargaining procedure to reach an allocation of the cooperative costs, players might not follow the reasoning of subadditivity which could lead to a coalition structure not containing the grand coalition (see Maschler 1992, p. 602, for a further discussion).

In the context of profit games (N, v) where payoffs are used instead of costs, the notion of *superadditivity* is used. Two disjoint subsets have an incentive to cooperate if the profits in case of cooperation are higher than the profits when acting alone:

$$v(S_1) + v(S_2) \leq v(S_1 \cup S_2) \qquad S_1, \ S_2 \subseteq N, \ S_1 \cap S_2 = \emptyset.$$

From a practical point of view, sub-/superadditivity can be interpreted as economies of scale or scope. Certainly, they should compensate negative implications arising in the cooperation. Derived from this subadditivity definition, we can define the *subadditive cover* $\hat{c}$ of the game c in coalitional form that satisfies

$$\hat{c}(S) = \min \left\{ \sum_{h=1}^{m} c(T_h) | \{T_1, \ldots, T_m\} \in \mathcal{P}(S) \right\} \quad \text{for all } S \subseteq N.$$

Thus, the total costs of a coalition S in the subadditive cover are the minimum costs that the coalition achieves when breaking up into a set of smaller disjoint coalitions (partition). This provides a possibility to transform any game in coalitional form into a corresponding subadditive game. Owen (2001) argues that the characteristic function assigns the maximin value to each $S \subseteq N$ (in case of profit games) of the two-person game played between S and $N \backslash S$. Hence, the property of superadditivity holds for every $|N|$-person game in characteristic function form (see Owen 2001, p. 213). When a cooperative game does not satisfy the property of sub-/superadditivity, cooperation would not be advisable.

Cooperative games can be divided into essential and inessential games. An *essential game* is characterized by

$$c(N) < \sum_{i \in N} c(\{i\}).$$

$c(N) \leq \sum_{i \in N} c(\{i\})$ follows directly from subadditivity but in the case of equality, the allocation problem would be trivial because every player would get exactly its stand-alone costs.

Furthermore, we can determine whether a cooperative game is *concave*:

$$c(S_1 \cup \{i\}) - c(S_1) \geq c(S_2 \cup \{i\}) - c(S_2) \qquad i \in N, \ S_1 \subset S_2 \subseteq N \backslash \{i\}.$$

The definition of concavity is related to subadditivity but much more severe. A concave characteristic function implies that a player causes a smaller increase in total costs while added to a subcoalition that contains more players. Some kind of "snowballing" effect occurs – the incentive for joining a coalition increases as the coalition grows. Fromen (2004, p. 87) argues that this behavior might be initiated by the network effect. Sometimes, an alternative formulation for concavity is used in the literature (see e.g., Shapley 1971):

$$c(S_1) + c(S_2) \geq c(S_1 \cup S_2) + c(S_1 \cap S_2) \qquad S_1, S_2 \subseteq N. \qquad (2.3)$$

Note that $c(\emptyset) = 0$ in combination with the concavity definition (2.3) implies subadditivity (2.2). Convex profit games follow the same interpretation:

$$v(S_1) + v(S_2) \leq v(S_1 \cup S_2) + v(S_1 \cap S_2) \qquad S_1, S_2 \subseteq N.$$

In the course of this thesis, we will deal with cooperative $|N|$-person TU games that are subadditive. Essentiality would be desirable but cannot be assured for practical problems. Nevertheless, in most cases it will be easy to check. Concavity would be a desirable property as well, especially from an analytical point of view. However, it limits the application possibilities too much. Starting in Chap. 5, we will introduce such games with special applications in supply chain management.

2.2.3 Variants and Fundamental Applications of the Classical Cooperative Game

This section presents variants of the before presented classical game that are also studied in the literature. Furthermore, we will discuss some fundamental applications of cooperative games that are well established in the literature and have their roots in operations research. The choice may be subjective and only wants to provide an overview of what is happening in the literature next to the later on presented applications and game variants.

Cooperative games with *restricted cooperation*, see Faigle (1989), are used to represent situations where only specific and not all coalitions $S \subseteq N$ are feasible. We will study such games in detail in Chap. 8.

Another variant are *bicooperative games* introduced by Bilbao (2000). Such games have non-orthogonal coalitions; i.e., the worth of a coalition depends on the actions of its complementary coalition. Bilbao et al. (2007, 2008) continue the study of bicooperative games while defining the solution concept of the core and the Shapley value for such games.

Lehrer (2002) extends the classical definition of a cooperative game by a *temporal aspect*: At each stage, a budget is distributed among the players. He shows that specific allocation processes converge to the core.

Multiple scenario cooperative games are studied by Hinojosa et al. (2005). Additionally, they extend the solution concepts of the core, the least core, and the nucleolus for those games. The costs of a coalition are valued in different scenarios in such games; i.e., the characteristic function is multidimensional.

Muto et al. (1989) introduce so-called *information market games*. They result from situations where intellectual property is used jointly. One player owns this property and the value of a coalition only depends on the presence of this player and its resource in the coalition.

Externality games are developed by Grafe et al. (1998). Here, each player's contribution to the value of the coalition consists of its own ability of contributing to the coalition and the second part arises from its simple presence in the coalition.

Owen (2001, p. 218) describes *weighted majority games* and *voting games* as special cases of *simple games*. Simple games have characteristic functions with values $\{0, 1\}$; i.e., a coalition either wins or loses. Closely related to voting games are *homogen games* discussed, for instance, by Maschler (1992, p. 628). Furthermore, Maschler (1992, p. 629) gives a survey of combinations of different games.

Fuzzy games represent situations where the membership of a player in coalition S can be stated in the interval $[0; 1]$ which is called participation level. The participation level of a classical game (Branzei et al. 2008a call them *crisp games*) is either 1 (if $i \in S$) or 0 (if $i \in N \setminus S$). Branzei et al. (2008a, p. 77) give an excellent survey of fuzzy games including fuzzy core variants and discussions on the convexity of fuzzy games.

Hsiao and Raghavan (1993) introduce *multichoice cooperative games* that are a generalization of a cooperative game in which the players have several activity levels at which each can choose to play. Hence, every player has an action space offering the possibility to do nothing or to work at a level k. These games were intensely studied by Hsiao and Raghavan (1992), van den Nouweland et al. (1995), Hsiao (1995a,b), Klijn et al. (1999), Calvo and Santos (2000), Peters and Zank (2005), Grabisch and Xie (2007), Hwang and Liao (2008), and Branzei et al. (2008a).

Several game variants got their names from a specific application or from classical operations research problems. Claus and Kleitman (1973) and Bird (1976) introduce *(minimum cost) spanning tree games*: The players are represented by nodes in a graph $G = (V, E)$ with the node set V (including a source) and $E = V \times V$ representing the edges. The graph has to be complete (every pair of nodes is connected via an edge) and the edges are undirected and weighted with costs or profits. The value of a coalition is determined by the minimal spanning tree connecting all players in the coalition and the source. Compare Borm et al. (2001, p. 152) and Curiel (1997, p. 129) for a general description and a literature survey regarding minimum cost spanning tree games.

Airport games are special network games (the fixed tree is a line graph) where the characteristic function values are determined by the "strongest" member in the coalition. The name of this game variant originates from its application: An airport operator needs to allocate the fixed capital costs for runway and terminal construction to the planes using the airport. These capital costs essentially depend on the largest plane that wants to land, therefore, the value (cost) for a coalition is determined by the largest plane/player in the coalition. These games are introduced by Littlechild and Thompson (1973) (see also Littlechild 1974; Littlechild and Thompson 1977a; Littlechild and Thompson 1977b; Dubey 1982; Potters and Sudhölter 1999; and Owen 2001, p. 334, for a detailed description and extensions).

Alike spanning network games, *flow games* are modeled in networks. Curiel (1997) describes flow games as a special form of *linear programming games* where the restrictions of a linear optimization problem depend on the coalitions. Moghaddam and Michelot (2009) present an algorithm for LP games that uses information from the final simplex tableau to allocate joint costs. Another well known variant of linear programming games are so-called *linear production games* introduced by Owen (1975). In such a game, the players try to maximize their profit via producing different products under the restriction of limited resources. Lower and upper bounds on the values of a linear production game are presented by Bjørndal and Jörnsten (2009). They determine the bounds via aggregating over the columns of the LP.

Another sort of games based on graphs is connected with the field of routing – *shortest path games*, *traveling salesman games*, and *Chinese postman games*. The main target of these games is to find a shortest path under specific conditions through a network and divide the costs or profits among the participating players along this path. Compare Borm et al. (2001, p. 155) and Curiel (1997, p. 111) for a broad survey. Estévez-Fernández et al. (2009) investigate routing games that take revenues into account as well. *Delivery games* which are based on Chinese postman games are studied by Hamers et al. (1999). The continuation of those games are vehicle routing games which are studied, for instance, by Göthe-Lundgren et al. (1996) and Engevall et al. (2004).

In *sequencing games* and *permutation games*, a number of jobs should be processed in a specific order that minimizes a cost criterion – the jobs represent the players in such a game. The difference between sequencing and permutation games is: If the members of coalition S are rearranged, it is allowed to jump over non-members in case of permutation games, but not in sequencing games (see Curiel et al. 1989 and Curiel 1997). *Assignment games* form a subclass of permutation games. In an assignment game, the player set N can be divided into two subsets such that the characteristic function value of a coalition is zero if it contains only players from one of the two subsets (see Curiel 1997, p. 55). Following Borm et al. (2001, p. 165), sequencing, permutation, and assignment games can be summarized under the topic of *scheduling games*.

2.2.4 Interval-Valued Games

We now present a special extension for the classical game that was introduced in Sect. 2.2.1. Classical cooperative game theory reaches its borders if the characteristic function values are not known with certainty. Interval-valued games are one concept to deal with uncertain values of the characteristic function, i.e., the players do not know the outcomes exactly. However, they know at least upper and lower bounds (with certainty) of the potential profits or costs generated by the coalitions, respectively. Hence, it is possible to make decisions based on all possible realizations which belong to the intervals. Formally, an interval-valued cost game is defined by a pair (N, c^{IV}). As before, N is the index set of players. $c^{IV} : 2^N \rightarrow I(\mathbb{R})$ is a characteristic cost function which assigns to every coalition $S \in 2^N$ a closed interval $c^{IV}(S) = [\underline{c}^{IV}(S); \overline{c}^{IV}(S)]$. The empty coalition receives $c^{IV}(\emptyset) = [0; 0]$. $I(\mathbb{R})$ is the set of all closed intervals in $\mathbb{R}$. Obviously, classical cooperative games are a special case of cooperative interval-valued games with $\underline{c}^{IV}(S) = \overline{c}^{IV}(S)$ for all coalitions $S \subseteq N$.

Cooperative interval-valued games are introduced by Branzei et al. (2003) where they apply this concept to bankruptcy situations. In such a situation, the estate is known with certainty, but for the claims exist only known intervals of real numbers. In addition to that, they develop two solution concepts related to the concept of the Shapley value. They continue their research in Branzei et al. (2004) for situations where the individual claims vary within closed intervals and in Branzei and Alparslan-Gök (2008) where they extend two allocation rules from classical bankruptcy games to the interval setting. Carpente et al. (2005) construct coalitional interval-valued games out of games in strategic form. The interval indicates the range bounded from below by the pessimistic prediction obtained using the lower value of the associated zero-sum game and bounded from above by the optimistic prediction obtained using the upper value of that game.

Theoretical results regarding interval-valued games are presented in several papers: Alparslan-Gök et al. (2009) investigate properties of two-person cooperative games with interval uncertainty – they study balancedness, superadditivity, and solution concepts like the core and the Shapley value (see Sect. 2.3). While looking at the corresponding classical cooperative games, which are selections of the interval-valued game, they extend the solution concept of the core (see also Sect. 2.3) for cooperative interval-valued games and define it as the union of the cores of all its selections. In Alparslan-Gök et al. (2008a), several solution concepts like the interval core, the interval dominance core, and stable sets are introduced as well as the notion of $\mathcal{I}$-balancedness. In a second paper, they focus on convex interval-valued games Alparslan-Gök et al. (2008b). Fundamental results regarding the relation between different solution concepts like the Shapley value, the Weber set, and the interval core are presented. Inspired by the classical big boss games (see Muto et al. 1988), Alparslan-Gök et al. (2008c) introduce big boss interval-valued games. Branzei et al. (2008b) continue the studies on big boss interval-valued games. How to deal with interval solutions in practical situations, is described by Branzei et al. (2008c).

On top of the application to bankruptcy games, there are further assignments to practical problems: For instance, Bauso and Timmer (2009) investigate a joint replenishment game (based on Hartman et al. 2000; Meca et al. 2003, 2004) where the demand is uncertain, but bounded by a minimum and maximum value. Furthermore, Alparslan-Gök et al. (2008d) extend the classical airport game, where costs for the runway need to be allocated, to an interval-valued game in which the costs of pieces of the runway are intervals. To summarize, Branzei et al. (2009) give a survey about cooperative interval-valued games including theoretical results as well as applications in economic and operations research situations.

2.3 Allocating Cooperative Costs

The fundamental question in cooperative game theory is how to allocate the costs of the grand coalition among the players. We already denoted a cost allocation with the vector $\pi = (\pi_1, \pi_2, \ldots, \pi_{|N|})$. π_i represents the portion of the total costs, incurred by the grand coalition N, which is allocated to player i. We will start our discussion with motivating cost allocation and a rough classification of allocation methods. As a basis for the later on defined solution concepts, we will explain desired properties of such cost allocations. Before we go into detail with game theoretical allocation approaches, we will shortly discuss the topic of cost allocation from an accounting viewpoint and show that there is a need for other allocation methods. We will concentrate our explanations regarding solution concepts on the concept of the core and its variants (including the nucleolus) as they are essential for the further content of this thesis. After this, we will shortly introduce the Shapley value in contrast to the core variants. For a more detailed description and discussion of other cooperative solution concepts like the bargaining set and the τ-value, compare, e.g., Myerson (1991, Sects. 9.4 and 9.6), Owen (2001, Chap. X and the following), Osborne and Rubinstein (1994, Part IV), Curiel (1997, Sects. 1.3 and the following), Fromen (2004, Chap. 4), or Peleg and Sudhölter (2007, Chaps. 3 and the following). For surveys on cost allocation in general see for instance, Shubik (1962), Young (1994), and Fromen (2004).

2.3.1 Motivation and Classification of Allocation Methods

Before discussing methods of cost allocation, we should analyze the motivation for allocating cooperative costs. Shubik (1962) is the first who studies cost allocation in combination with game theory. He argues that cost allocation, thus the individual cost shares, can guide individual decisions in such a way that they are preferable for the cooperation. Hence, he develops an incentive system for decentralized control (Shubik 1962, pp. 328 and 333). Moriarty (1981a) displays motives for cost allocation while dividing them in three groups: textbook rationales (e.g., external reporting of inventory values or for cost control and product pricing), rationales presented in the recent literature (encourage cooperation and discourage wasteful consumption), and rationales involving decision making (signaling optimal capacity adjustments and relative profitability of products). Demski (1981, p. 152) calls the three most important motives decomposition (decomposing complex decision problems), motivation (cooperating partners should act in the interest of the grand coalition), and coordination (coordinate actions of several partners to act in the interest of the grand coalition). Nagarajan and Sošić (2008, p. 2) mention stability of the cooperation as the most important strategic reason for allocating costs, especially in interorganizational cooperations (see Sect. 4.1).

Barton (1988) uses a very obvious classification method for allocation concepts: *complex game theoretic approaches* and *practical accounting type solutions* – the former category uses methods from cooperative game theory as described in Sects. 2.3.4–2.3.9, the latter uses proportional values (see Sect. 2.3.3). In contrast to game theoretical approaches, the second class counts for understandability and practical applicability because game theoretical concepts have a higher complexity. Such proportional values are also called activity measures and are based, for instance, on quantity or sales figures (see Biddle and Steinberg 1984, p. 18, and Arcelus et al. 1997, p. 446). Biddle and Steinberg (1984, p. 18) distinguish a third class with *normative allocation proposals*. The normative concepts should provide a general measure for allocation but no general definition exists for this class of concepts.

Fromen (2004) develops a more sophisticated classification of the existing solution methods where concepts are sorted in four classes: "classical" concepts, bargaining concepts, proportional concepts, and others (see Fromen 2004, p. 96).

The first application of cooperative game theory can be found in Ransmeier (1942) where costs for a dam system should be allocated among the participatory users (see also Straffin and Heaney 1981 and Young et al. 1982). The work of Moriarity (1981b) presents several papers dealing with cost allocation from an accounting viewpoint, but integrates game theory methods as well. Contrary to the field of non-cooperative game theory, where the Nash equilibrium is the substantial solution method, many different solution concepts exist in cooperative game theory where no "single, all-purpose solution" is available (see Young 1994, p. 1230). Choosing the right method heavily depends on the context, organizational goals, amount of available information, etc.

2.3.2 Properties of Cost Allocations

There needs to be an incentive to cooperate, otherwise, cooperation will not occur and the outcome will be inefficient. One of the first authors discussing the importance of "fair" allocation techniques was Lemaire (1984). He describes a practical example where common accounting methods based on activity measures to allocate joint costs lead to plenty of problems and wrong incentives. Based on the explanations of Lemaire (1984) and Myerson (1991), we will now discuss important properties a cost allocation should achieve.

Theoretically, the set of solutions could contain all vectors π where at least the total costs are allocated:

$$\sum_{i \in N} \pi_i \geq c(N).$$

We call such solutions *feasible allocation vectors*. Nevertheless, the total costs of the grand coalition should be allocated completely (neither more nor less) because if we allocate more than $c(N)$, the share of at least one player could be improved (decreased) without harming another player. *Efficiency* assures that the sum of the cost shares equals the total costs $c(N)$ of the grand coalition N:

$$\sum_{i \in N} \pi_i = c(N). \tag{2.4}$$

In some references, the terms *feasibility* or *group rationality* are used instead of efficiency (e.g., Myerson 1991, p. 427, and Maschler 1992, p. 595). Maschler (1992) calls an allocation vector that satisfies (2.4) a *preimputation*.

As described at the beginning of this chapter, cooperative game theory assumes that players act rational. Thus, it is a desirable property for a cost allocation that a single player has lower (or at least equal) costs while cooperating than when acting alone. This property is called *individual rationality*:

$$\pi_i \leq c(\{i\}) \qquad i \in N. \tag{2.5}$$

An allocation is called an *imputation* if it secures efficiency (2.4) and individual rationality (2.5). No player would accept an allocation that is no imputation because he would be better off not cooperating. Therefore, most of the solution concepts in cooperative game theory are based upon the set of imputations.

We can further extent the idea of individual rationality because not only single players act rational, but also subcoalitions. A coalition $S \subset N$, $S \neq \emptyset$ of players will cooperate with the remaining players $N \backslash S$ if S cannot improve on the allocation; i.e., *coalitional rationality* holds

$$\sum_{i \in S} \pi_i \leq c(S) \qquad S \subseteq N, \ S \neq \emptyset. \tag{2.6}$$

With constraints (2.5) and (2.6), it is possible to assure that no member of the coalition has any incentive to leave the grand coalition and form a smaller coalition S – we also speak of a *stable* allocation because a cooperation of the grand coalition would not stand in the long run if a partner would be better off not cooperating in the grand coalition. Note, restrictions (2.4) and (2.5) are special cases of (2.6) when S are the singletons or N, respectively (efficiency requires an inequality in the other direction). Instead of rationality constraint, the notion *stability* constraint is used equivalently. The principles of individual and coalitional rationality are already stated in Ransmeier (1942, p. 220).

Lemaire (1984) states that no player should be charged less than its marginal costs and called this property *collective rationality* or *marginality principle*:

$$\pi_i \geq c(N) - c(N \backslash \{i\}) \qquad i \in N.$$

If a player pays less than its marginal costs $c(N) - c(N \backslash \{i\})$, then it is effectively subsidized by other players. But there is no need to continue with this restriction because it follows immediately from

Efficiency (2.4):   Stability (2.6):

$$\sum_{i \in N} \pi_i = c(N)$$ $$\sum_{i \in S} \pi_i \leq c(S)$$

$$\rightarrow \sum_{j \in N\setminus\{i\}} \pi_j + \pi_i = c(N)$$ $$\rightarrow \sum_{j \in N\setminus\{i\}} \pi_j \leq c(N\setminus\{i\})$$

$$\rightarrow \pi_i = c(N) - \sum_{j \in N\setminus\{i\}} \pi_j$$

$$\rightarrow \pi_i \geq c(N) - c(N\setminus\{i\})$$

Young et al. (1982) formulate the marginality principle also for coalitions which can be derived from (2.4) and (2.6) as well:

$$\sum_{i \in S} \pi_i \geq c(N) - c(N\setminus S) \qquad S \subset N.$$

A notion coming together with desired properties of a cost allocation is *fairness* (or *equity*). Due to the wide meaning of this term, Young et al. (1982, p. 464) suggest a cost allocation where the participants agree in principle to the proposed allocation. Hence, Young et al. (1982) call an allocation fair if it fulfills (2.4) and (2.6) because it constitutes the minimum incentive to join the grand coalition for every partner. We will pick up the discussion regarding fairness later on, in particular in Sects. 2.3.6 and 3.4. For further properties that are very specific and not used in the following discussion, see e.g., Lemaire (1984), Young (1994), and Zelewski (2007).

   In the following sections, we will present different solution concepts which might assure the discussed properties. A *solution concept* is a rule for allocating the total costs of the grand coalition while a *solution* gives a concrete allocation vector π. Solution concepts can be distinguished into those that yield a solution containing exactly one allocation vector (e.g., the Shapley value) and those that give a set of allocation vectors (e.g., the core).

2.3.3 Non-Game-Theoretical Cost Allocation Methods

The problem of allocating joint or common costs has its roots in the field of accounting. For a distinction between joint and common costs see Biddle and Steinberg (1984, p. 4). In the following, it is sufficient to use the term joint costs and neglect the distinction: *Joint costs* are costs for which it is impossible or would be too complex to incur the costs relating to several cost objects individually (Biddle and Steinberg 1984, p. 4).

   Furthermore, Biddle and Steinberg (1984, p. 3) give an abstract definition for cost allocation: *Cost allocation* is an "efficient partitioning of a cost among a set of cost objects". Note that in the world of accounting, a cost object is a product in most of the cases. For contributions concerning cost allocation from an accounting viewpoint, see e.g., Pfaff (1994) and Snyder and Davenport (1997). Pfaff (1994) investigates disadvantages when joint costs are not allocated and fully allocated, respectively, and suggests an allocation mechanism which is optimal under specific conditions. A more practical discussion is presented by Snyder and Davenport (1997).

Transfer prices are related to the topic of cost allocation: Biddle and Steinberg (1984, p. 5) state that *"transfer prices* serve to allocate revenues against which allocated costs are matched". Compare also Illner (1999, p. 21) for a detailed discussion regarding the classification of cost allocation and transfer prices.

Certainly, one of the easiest approaches to allocate costs is the *equal amount method*. As the name implies, total costs of the cooperation will be allocated in equal parts to the partners. Apparently, the input level of one partner is ignored completely. The *proportional method* uses different information about the partners in the cooperation to allocate the costs; e.g., purchasing volume, individual costs, or the willingness to pay of every partner (Young et al. 1982, p. 467, call this the *separable costs-remaining benefits method*). These concepts are working with *activity measures*. Another group of practical allocation methods is based on marginal costs: *equal or proportional allocation of the non-marginal costs*. The *equal allocation of the total gain method* employs information regarding the cost savings for allocation. A *serial cost sharing* is proposed by Moulin and Shenker (1992) where an amount is allocated to each partner depending on increasing or decreasing demand, respectively. This concept is not based on game theory but the authors show that it yields a unique Nash equilibrium. Hence, the idea of serial cost sharing finds its way into game-theoretical research, e.g., de Frutos (1998) and Albizuri et al. (2003). Compare, e.g., Lemaire (1984, p. 65), Heijboer (2003, p. 70), and Schotanus (2007, p. 163), for more details regarding the aforementioned concepts.

Normative allocation concepts are introduced in Moriarity (1975) with the main target to avoid the allocating problems when using activity measures in practical situations. Moriarity's idea is to integrate cost savings that are yielded by the cooperation. Louderback (1976) criticizes the approach of Moriarity because it cannot be avoided in general that partners will be subsidized. Thus, he integrates inside and outside incremental costs caused by the cooperation. Balachandran and Ramakrishnan (1981) combine the two approaches with their propensity-to-contribute concept. Joint costs will be allocated according to each partner's willingness to pay more than a minimum value. The concept of Gangolly (1981) is based on Moriarity's allocation scheme. In contrast, Gangolly (1981) considers not only single players that benefit from cost savings in the grand coalition, but also subcoalitions.

For a comparison of game theoretical approaches and practical oriented solution concepts see e.g., Nagarajan et al. (2008, p. 10), Schotanus (2007, pp. 141 and 169), and Heijboer (2002, p. 282). Compare, furthermore, Fromen (2004, p. 43) for a general discussion of non-game-theoretical methods for cost allocation.

Young et al. (1982, p. 465) and Lemaire (1984) explain that most of the common accounting methods presented above fail to satisfy natural requirements. With the help of a practical example, Lemaire (1984, p. 64) shows that these accounting methods have to be rejected because none of them meets the rationality restrictions. In addition to that proof, Lemaire (1984, p. 77) reviews successful applications of game theoretic approaches for solving practical cost allocation problems. However, he also describes a big drawback because game theoretic methods need more information in terms of the $2^{|N|} - 1$ cost values $c(S)$ (see also Young et al. 1982, p. 471). We will pick up this criticism in Sect. 3.2 where we will introduce a procedure to compute cost allocations which does not need explicit information for all subcoalitions $S \subset N$. Beforehand, we will present game theoretical methods to solve cost allocation problems.

2.3.4 The Core

The core is an essential and one of the most prominent solution concepts to allocate costs (or profits) in problems of cooperative game theory, especially in economic theory. It combines the first three properties mentioned in Sect. 2.3.2 (2.4)–(2.6). The notion of the core is credited to Gillies (1959). However, the idea was prefigured in the early theoretical literature on cost allocation and already discussed by Edgeworth (1881). The core is based on a very simple idea: An imputation is called stable if no coalition could form to reach a better result than in the grand coalition. This leads to the before presented property of coalitional rationality (2.6). Hence, the core combines the property of efficiency with individual and coalitional rationality.

In the literature, the core is mostly defined with the help of the term domination: The core $C(N, c)$ of a cooperative game with $|N|$ players and the characteristic cost function c is the set of all non-dominated imputations (see e.g., Owen 2001, p. 218). Following for instance Owen (2001, p. 215) the imputation π^1 dominates the imputation π^2 through the coalition S if the two following constraints hold:

1. $\pi_i^1 < \pi_i^2$ for all $i \in S$
2. $\sum_{i \in S} \pi_i^1 \geq c(S)$

π^1 dominates π^2 if there is at least one coalition S such that π^1 dominates π^2 through S. The first condition assures that every player in S is preferring π^1 to π^2. The second restriction claims that the coalition S does not achieve less costs than they should bear while working alone without the players in the grand coalition – it assures feasibility. Hence, non-dominated imputations possess some kind of stability which directly leads to the definition of the core:

$$C(N, c) = \left\{ \pi \in \mathbb{R}^{|N|} \, \middle| \, \underbrace{\sum_{i \in N} \pi_i = c(N)}_{\text{efficiency}} \text{ and } \underbrace{\sum_{i \in S} \pi_i \leq c(S)}_{\text{rationality/stability}} \text{ for all } S \subset N, \, S \neq \emptyset \right\}.$$

$$(2.7)$$

The efficiency condition in combination with the rationality condition with $S = \{i\}$ leads to the definition of an imputation. Furthermore, assume that π^1 satisfies both conditions and that $\pi^2 < \pi^1$ for all $i \in S$. In combination with the rationality constraints, we would get $\sum_{i \in S} \pi_i^2 < c(S)$. Hence, it is not possible that π^2 dominates π^1 (see Owen 2001, p. 218, for a further discussion).

Thus, a core allocation assures that every player is better off in the grand coalition. Otherwise, there would be some coalition S where the players in S could all do strictly better than in the grand coalition while cooperating without the players $N \setminus S$ and dividing $c(S)$ among themselves. Additionally, this is adequate to the notion of pareto optimality because there is no imputation outside the core where one player could get a better result without penalizing another player; i.e., no player or coalition will be subsidized. For a detailed discussion and derivation, see Myerson (1991, p. 427) and Owen (2001, p. 218).

Myerson (1991) describes the core as very appealing because it treats players symmetrically, includes pareto-efficient allocations, and reflects weakness or power of the players according to the characteristic function.

The efficiency constraint (2.4) defines a hyperplane H_N in R^N. Obviously, the core is a subset of this hyperplane H_N and, due to the individual rationality constraints (2.5), the core is bounded. Hence, the core is a compact convex polyhedron of at most $(|N| - 1)$ dimensions (see e.g., Shapley 1971 and Owen 2001, p. 219). That generally means, the core has more than one point; i.e., more than one outcome (allocation vector) is stable. If we know at least two core elements, we can reason, due to the convexity, that the core contains an infinite number of imputations. Hence, for choosing exactly one core element, we need some more rules. Like in Maschler et al. (1979), the core for a three player game can be illustrated graphically in a triangle where each point displays a suggested cost allocation (π_1, π_2, π_3) such that the efficiency constraint holds (see Fig. 2.1). The gray area in the center of the triangle displays the core.

Basically, values $\pi_i < 0$ are allowed. Nevertheless, all cost assignments in the core being non-negative ($\pi_i \geq 0$) holds automatically if the characteristic function is monotone; i.e., $c(S_1) \leq c(S_2)$ for all $S_1 \subseteq S_2 \subseteq N$. Compare Drechsel and Kimms (2010c) for the lemma and proof:

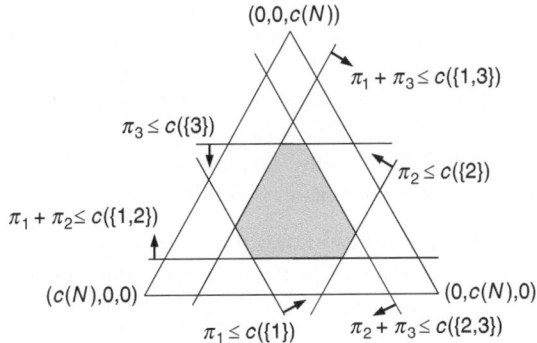

Fig. 2.1 Graphical illustration of the core for a game with three players

Table 2.1 Two numerical examples with nonempty and empty core

	$c(\{1\})$	$c(\{2\})$	$c(\{3\})$	$c(\{1,2\})$	$c(\{1,3\})$	$c(\{2,3\})$	$c(\{1,2,3\})$
Example 1	5	5	5	7	7	7	10
Example 2	5	5	5	7	7	7	11

$$\pi_i = c(N) - \sum_{j \in N \setminus \{i\}} \pi_j \geq c(N) - c(N \setminus \{i\}) \geq 0 \qquad \text{for all } i \in N.$$

The concept of the core has its drawbacks, particularly, the possible emptiness of the core. Table 2.1 presents two small examples: Both games are subadditive (see Sect. 2.2.2), however, Example 1 has a nonempty core, whereas Example 2 has an empty core. Compare the resulting core constraints:

Example 1:   Example 2:

$$\pi_1 + \pi_2 + \pi_3 = 10 \qquad\qquad \pi_1 + \pi_2 + \pi_3 = 11$$
$$\pi_1 \leq 5 \qquad\qquad \pi_1 \leq 5$$
$$\pi_2 \leq 5 \qquad\qquad \pi_2 \leq 5$$
$$\pi_3 \leq 5 \qquad\qquad \pi_3 \leq 5$$
$$\pi_1 + \pi_2 \leq 7^* \qquad\qquad \pi_1 + \pi_2 \leq 7^*$$
$$\pi_1 + \pi_3 \leq 7^* \qquad\qquad \pi_1 + \pi_3 \leq 7^*$$
$$\pi_2 + \pi_3 \leq 7^* \qquad\qquad \pi_2 + \pi_3 \leq 7^*$$

The three *-marked constraints lead to $\pi_1 + \pi_2 + \pi_3 \leq 10.5$, hence, for $c(N) = 11$ the core is empty. For the second example, no matter what allocation may ultimately occur, there is always a coalition that has an incentive to leave the grand coalition. Though, the greater the economies of scale, the more likely a core allocation can be found. In general, the core of a subadditive three player game is not empty if $c(\{1,2\}) + c(\{1,3\}) + c(\{2,3\}) \geq 2c(N)$ holds.

Having a game with an empty core does not imply that the players could not agree on cooperation and an allocation. There might be binding agreements that could not be broken without consent of the other players. On the other hand, if a coalition exists, that would be better off acting alone, it might hesitate to speak against the allocation because it fears further negotiations where the result is worse than in the beginning. Myerson (1991) formulates three assumptions under which a coalition might think about negotiating for a better allocation:

1. Prior agreements do not forbid such renegotiation.
2. The new agreement would be final – no more bargaining rounds follow.
3. In case of failing agreements, the allocation, valid before starting the negotiation, will be still valid.

Compare Myerson (1991, p. 431) for a further discussion.

Shapley (1971) analyzes the core of convex profit games: He proves that the core of a convex profit game is not empty. Furthermore, he studies the geometry of the core and establishes that the set of marginal worth vectors is precisely the set of vertices of the core of a convex game. It is shown by Weber (1978) that the convex hull of the set of marginal worth vectors contains the core of any game. Ichiishi (1981) relates the study of convex profit games and their marginal vectors to a special greedy algorithm for linear programming. The results can be transferred to concave cost games: A *marginal worth vector* yields a core element for a concave cost game. Let σ be a permutation on N. Define

$$P_j^\sigma \equiv \{i \in N | \sigma(i) < \sigma(j)\}$$

as the set of members in N which precede j with respect to the order σ. Hence, the marginal worth vector for a given permutation is defined as the vector

$$a^\sigma(c) \in \mathbb{R}^{|N|} \quad \text{with} \quad a_j^\sigma(c) = c(P_j^\sigma \cup \{j\}) - c(P_j^\sigma) \quad \text{for every } j \in N.$$

However, there exist games that are not concave and for many games it is not clear whether or not the core is empty. Consequently, many contributions to cooperative games concentrate on proving the non-emptiness of the core for specific games. Often, they do not compute core elements explicitly but they prove the balancedness of the game by using the *Bondareva–Shapley Theorem* (see Bondareva 1963 and Shapley 1967) that is based on duality theory of linear programming: The collection $\{S_1, \ldots, S_k\}$ of coalitions of N is called balanced if there exist positive numbers $\lambda_1, \ldots, \lambda_k$ such that

$$\sum_{j; S_j \ni i} \lambda_j = 1 \quad \text{for every } i \in N.$$

$\lambda_1, \ldots, \lambda_k$ are called balancing weights.

The Bondareva–Shapley Theorem signifies that the core of the profit game v is nonempty if and only if for every minimal balanced collection $\{S_1, \ldots, S_k\}$ with balancing weights $\lambda_1, \ldots, \lambda_k$ the inequality $\sum_{j=1}^k \lambda_j v(S_j) \leq v(N)$ holds (see Bondareva 1963 and Shapley 1967 for the proof).

Kannai (1999) gives a broad survey on balanced games including TU and NTU games. Apart from the non-emptiness of the core, another part of literature deals with proving concavity or convexity of specific games, respectively, and other core properties. González-Díaz and Sánchez-Rodríguez (2007) discuss a particular selection from the core. They call it the core-center because this allocation rule selects a centrally located point within the core if the core is not empty. They continue the investigation of the core-center in González-Díaz and Sánchez-Rodríguez (2009). The studies about marginal vectors and convexity of profit games are advanced by van Velzen et al. (2004). They use combinatorial arguments to obtain sets of marginal vectors that characterize convexity of a game and display a formula for the minimum cardinality of sets of those marginal vectors. Hamers et al. (2002)

deal with marginal vectors as well and prove that the extreme points of the core of assignment games are marginal vectors. The properties of the core's extreme points are studied by Núñez and Rafels (1998). They prove that the extreme points have the reduced game property with respect to the Davis and Maschler reduced game (see Davis and Maschler 1965), but not the converse reduced game property. Necessary and sufficient conditions under which the core coincides with the bargaining set for superadditive profit games are proved by Solymosi (1999). Shapley and Shubik (1971) present an extensive discussion of the core regarding assignment games which gives among other things insights into the geometric structure of the core.

Due to the described drawbacks regarding existence and uniqueness of the core, some authors argue that the core is only little applicable to practical allocation problems, but should be used to categorize the set of imputations and evaluate other solution concepts (see e.g., Fromen 2004, p. 100).

In papers where core elements are computed explicitly, very specific approaches are constructed that depend highly on the underlying problem and cannot be generalized. We will discuss the computation of core elements in Chap. 3 in detail.

As a reaction to the described drawbacks, several related concepts has been developed in the literature to overcome the difficulties of nonexistence and non-uniqueness. These concepts are based on strengthening or relaxing the stability constraints that define the core.

The following presented core variants are classified according to Faigle et al. (1998b) into two groups: At first, we introduce core variants where the characteristic function values are varied by an absolute amount – *additive core variants*. The second class of core variants is marked by a relative variation of the characteristic function values – *multiplicative core variants*. Afterwards, we will introduce another two core variants which do not fit in this classification: a core variant specifically developed for non-monotone cooperative games and the interval core for the before presented interval-valued games (see Sect. 2.2.4).

2.3.5 Additive Core Variants

The ϵ-Core

The ϵ-core is introduced by Shapley and Shubik (1966). They distinguish between the *strong ϵ-core*

$$C_\epsilon(N, c) = \left\{ \pi \in \mathbb{R}^{|N|} \,\middle|\, \sum_{i \in N} \pi_i = c(N) \text{ and } \sum_{i \in S} \pi_i \leq c(S) + \epsilon \right.$$
$$\left. \text{for all } S \subset N,\ S \neq \emptyset \right\}.$$

and the rarely used *weak ϵ-core*

$$C_\epsilon^{weak}(N,c) = \left\{ \pi \in \mathbb{R}^{|N|} \middle| \sum_{i \in N} \pi_i = c(N) \text{ and } \sum_{i \in S} \pi_i \leq c(S) + |S|\epsilon \right.$$

$$\left. \text{for all } S \subset N, \ S \neq \emptyset \right\}.$$

ϵ is a given parameter used to enlarge the core by a small amount. Shapley and Shubik (1966) call these two variants *quasi-cores*. Faigle et al. (1998b) use this formulation under the name *additive ϵ-core*. The above formulation provides a possibility to take the costs of coalition formation into account – a fixed charge ϵ. Thus, ϵ can be interpreted as threshold for a blocking maneuver resulting from a stability constraint. In other words, if founding a coalition $S \subset N$ while quitting the grand coalition causes a cost of $\epsilon > 0$, then the grand coalition is stable even if the coalition S receives a cost share larger than $c(S)$ as long as $\sum_{i \in S} \pi_i \leq c(S) + \epsilon$ is fulfilled. The same holds if a third party offers a reward $\epsilon < 0$ for forming a coalition $S \subset N$ and leaving the grand coalition. In this case, the grand coalition is stable as long as the sum of the cost shares for the players in coalition S is smaller or equal to the coalition cost $c(S)$ less the reward.

As already mentioned, the core of a cooperative game might be empty. Obviously, if we choose a sufficiently large ϵ, the ϵ-core of the same instance is not empty. Otherwise, the ϵ-core lies in the core with an $\epsilon \leq 0$ if the core of an instance is not empty; i.e.,

$$C_\epsilon^{weak}(N,c) \supset C_\epsilon(N,c) \supset C(N,c).$$

Evidently, the core is a special case of the ϵ-core; i.e., $C_0(N,c) = C(N,c)$. In the following, we will simply use the term ϵ-core when meaning the strong ϵ-core.

Fromen (2004) argues that the ϵ-core is actually not a self-contained solution concept, but should be used particularly to analyze games with an empty core.

The Least Core

Continuing the approach of the ϵ-core, Maschler et al. (1979) develop the so-called least core $C^L(N,c)$ to ensure existence and uniqueness of the allocation vector. They define the least core as the intersection of all nonempty ϵ-cores. Hence, the least core is the ϵ-core with the smallest possible $\epsilon \in \mathbb{R}$ such that $C_\epsilon(N,c)$ is not empty:

$$C^L(N,c) = \bigcap_{\epsilon \in \mathbb{R}: \ C_\epsilon(N,c) \neq \emptyset} C_\epsilon(N,c).$$

If the core of a game is not empty, then $\epsilon \leq 0$ and the least core is centrally located within the core. By contrast, if the core is empty ($\epsilon > 0$), then the least core can be seen as revealing the "latent" position of the core. The least core can be used to define the problem of proving non-emptiness of the core as an optimization problem that minimizes ϵ under the condition that $C_\epsilon(N, c)$ is not empty. As already mentioned, an objective function value $\epsilon^* \leq 0$ indicates that the core is not empty and the corresponding set $C_{\epsilon^*}(N, c)$ is the least core. For a nonempty core ($\epsilon^* \leq 0$), the following inclusion is straightforward:

$$C_{\epsilon^*}(N, c) = C^L(N, c) \subseteq C(N, c).$$

Obviously, the notion of the least core can be used to define the *weak least core* – compare the definition of the weak ϵ-core. For a detailed discussion and computation of the least core, see Schulz and Uhan (2007).

The Nucleolus

Schmeidler (1969) introduces the *nucleolus*. This concept is based on the idea of maximizing lexicographically the minimal satisfaction of each coalition step by step. Maschler et al. (1979) and Maschler (1992, p. 611) give illustrative descriptions to motivate the concept of the nucleolus. The measure for satisfaction is the so-called excess:

$$e(S, \pi) = c(S) - \sum_{i \in S} \pi_i.$$

Hence, the excess results from the difference between stand-alone costs (independent from the players $j \in N \backslash S$) and the costs the players $i \in S$ have to bear when cooperating in the grand coalition. The excess for the grand coalition and the empty coalition is zero, respectively:

$$e(\emptyset, \pi) = c(\emptyset) - \sum_{i \in \emptyset} \pi_i = 0$$

$$e(N, \pi) = c(N) - \sum_{i \in N} \pi_i = 0.$$

There are many variants of the nucleolus which mostly differ on the definition of the excess. A stepwise computation means that starting with the set of imputations, we first determine an imputation that maximizes the minimal excess. In the second step, an imputation that maximizes the second smallest excess is determined and so on. The procedure stops latest if all coalitions are considered.

Maschler (1992, p. 610) and Owen (2001, p. 322) give formal derivations of the nucleolus. Let $\theta(\pi)$ be a vector containing the excesses of all coalitions arranged in non-decreasing order

$$\theta(\pi) := (e(S_1, \pi), e(S_2, \pi), \ldots, e(S_{2^{|N|}}, \pi)).$$

Note, this order depends on the allocation vector π and does not need to be unique. $\theta(\pi^1)$ is called *lexicographically smaller* than $\theta(\pi^2)$ if a positive integer q exists such that $\theta_i(\pi^1) = \theta_i(\pi^2)$ whenever $i < q$ and $\theta_q(\pi^1) < \theta_q(\pi^2)$. We denote this with $\theta(\pi^1) \prec \theta(\pi^2)$. Thus, the nucleolus can be defined as

$$\mathcal{N}(N, c) = \left\{\pi^1 \in I(c) \,\middle|\, \theta(\pi^1) \succeq \theta(\pi^2) \text{ for all } \pi^2 \in I(c)\right\}$$

with $I(c)$ being the set of imputations. Schmeidler (1969) proves that the nucleolus is inside the core if the core is not empty (see also Maschler 1992). That is one reason why the nucleolus is willingly used. Provided that the core exists, the nucleolus yields a unique allocation in the core. Furthermore, the nucleolus always exists. Shubik (1982) states that the nucleolus shows the location of the "latent position" of the core if the core is empty and the core's center if it is not empty. The second statement is difficult to confirm: Maschler et al. (1979) provide two numerical examples with the same set of imputations and the same core but they have different values for the nucleolus. See Maschler et al. (1979) for an explanation regarding the geometrical position of the nucleolus inside the core.

2.3.6 Multiplicative Core Variants

The Approximate Core

In case of multiplicative core variants, there are contributions regarding the so-called *approximate core* or α-*core* (see e.g., Jain and Mahdian 2007, p. 389, and Schulz and Uhan 2007, p. 2). This concept allows inefficient solutions. The formal definition is

$$C_\alpha(N, c) = \left\{\pi \in \mathbb{R}^{|N|} \,\middle|\, \alpha c(N) \leq \sum_{i \in N} \pi_i \text{ and } \sum_{i \in S} \pi_i \leq c(S) \text{ for all } S \subseteq N, \, S \neq \emptyset\right\}.$$

$$(2.8)$$

The problem of proving the non-emptiness of the core can be formulated, using the notion of the approximate core, as an optimization problem that maximizes α under the restriction that $C_\alpha(N, c)$ is not empty; i.e., that the efficiency and stability constraints following (2.8) hold. If this optimization problem yields an optimum objective function value $\alpha^* = 1$, then the core is not empty; i.e., if $\alpha^* = 1$ then

$$C_{\alpha^*}(N, c) = C(N, c).$$

The Minmax Core

Allocations in the core or the approximate core can be considered as stable because no player has an incentive to break off the grand coalition. And from this point of view, they can be seen as fair in a weak sense. But those allocations might not be considered as inherently fair in a strong sense because players or subcoalitions in the grand coalition benefit more than others from the cost decrease. We can define benefit as the deviation of $\sum_{i \in S} \pi_i$ from $c(S)$. The ϵ-core and the least core guarantee that the absolute benefit for each coalition is at least ϵ, hence, they can be seen as fair core allocations as long as $\epsilon \leq 0$. However, if the $c(S)$ values differ a lot among the coalitions S, absolute values may not be regarded as fair. For practical use, we advise to use relative benefits. Compare Drechsel and Kimms (2010a) for some ideas of implementing fairness aspects into the notion of the classical core – these ideas will be discussed in Sect. 3.4 as well. Furthermore, Frisk et al. (2010) develop a relative measure which they call the *equal profit method*. They minimize the difference in relative savings between two participants i and j:

$$\frac{\pi_i}{c(\{i\})} - \frac{\pi_j}{c(\{j\})}.$$

However, all of these ideas consider fairness among the individual players only; i.e., the benefit of subcoalitions containing two and more players are not taken into account. Therefore, Drechsel and Kimms (2010b) propose another concept which they call the *minmax core*. The minmax core can be used to determine efficient, rational, and fair cost allocations. The principle of fairness is tackled by means of the minmax-principle. The benefit of a subcoalition S is measured as the relative benefit in percentages of $c(S)$. That means, the lower the cost assigned to a subcoalition S, the greater the benefit. The subcoalition that receives the lowest benefit (i.e., highest cost assignment in percentages) determines the fairness measure. To develop the mathematical formulation of the minmax core, we use the multiplicative ϵ-core of Faigle and Kern (1993) and Faigle et al. (1998b) (in several other papers it is called ϵ-approximation of the core or ϵ-core, see e.g., Faigle et al. 1998a; Faigle and Kern 1998). Let $(1 - \epsilon) = \eta$:

$$C_\eta(N, c) = \left\{ \pi \in \mathbb{R}^{|N|} \,\middle|\, \sum_{i \in N} \pi_i = c(N) \text{ and } \sum_{i \in S} \pi_i \leq \eta\, c(S) \text{ for all } S \subset N,\ S \neq \emptyset \right\}.$$

Drechsel and Kimms (2010b) call this the *η-core*. η assures that to no coalition S a cost share greater than η percentage of its stand-alone cost $c(S)$ is assigned – provided that $C_\eta(N, c)$ is not empty. This core variant suits all applications where coalition costs are changed proportional to the value (e.g., a sales tax). Faigle et al. (1998b) compare it to other "taxation models" in the literature like the weak ϵ-core defined above or the modification by Tijs and Driessen (1986) for multiplicative ϵ-tax games:

$$C_\epsilon^{tax}(N, c) = \left\{ \pi \in \mathbb{R}^{|N|} \,\middle|\, \sum_{i \in N} \pi_i = c(N) \text{ and} \right.$$

$$\left. \sum_{i \in S} \pi_i \leq c(S) + \epsilon \left(\sum_{i \in S} c(\{i\}) - c(S) \right) \text{ for all } S \subset N,\ S \neq \emptyset \right\}.$$

Faigle et al. (1998b) state that this core variant equals their multiplicative ϵ-core if the game is zero-normalized (i.e., $c(\{i\}) = 0$ for all $i \in N$). Faigle et al. (1998b) do not claim to rank the different core variants but show the value of the concept of the multiplicative ϵ-core with some practical examples.

As with the other core variants before, we can use the η-core to prove non-emptiness of the core via stating it as an optimization problem which minimizes η under the condition that $C_\eta(N, c)$ is not empty. If the optimum objective function value is $\eta^* \leq 1$, then the core is not empty (the reverse might not hold for $c(S) = 0$).

Drechsel and Kimms (2010b) call $C_{\eta^*}(N, c)$ the *minmax core* $C_M(N, c)$. It is straightforward that $C_{\eta^*}(N, c) \subseteq C(N, c)$ if the core is not empty. The minmax core can be interpreted as containing cost allocations where the worst benefit over all subcoalitions is as good as possible.

We already mentioned that $\eta^* \leq 1$ indicates non-emptiness of the core. However, even though the core is not empty, η^* could be greater than 1. This might happen if a coalition $S \subseteq N$ ($S \neq \emptyset$) exists with $c(S) = 0$. To make this more obvious, assume a coalition S with $c(S) = 0$. An efficient allocation requires

$$\sum_{i \in S} \pi_i + \sum_{i \in N \setminus S} \pi_i = c(N).$$

Two of the minmax core defining stability constraints are

$$\sum_{i \in S} \pi_i \leq 0 \quad \text{and} \quad \sum_{i \in N \setminus S} \pi_i \leq \eta \, c(N \setminus S).$$

Consequently, we have

$$\sum_{i \in N \setminus S} \pi_i \geq c(N).$$

Hence, $c(N) \leq \eta \, c(N \setminus S)$ which may enforce $\eta^* > 1$. Note that this is not necessarily true, in particular, if c is not monotone.

The idea of using such relative measures has been used by only few authors so far: Apart from the above mentioned references, there is Young et al. (1982) who introduce the so-called *proportional least core* for a profit game by imposing a minimum tax t (or subsidy) on all coalitions in proportion to their cost so that the proportional least core is not empty. The mathematical definition is nearly the same as for the minmax core.

The Proportional Nucleolus

Drechsel and Kimms (2010b) remark that the idea of the minmax core can be straightforwardly enhanced: In a first step, the worst benefit shall be as good as possible. In a second step, the second worst benefit (among those solutions computed in the first step) shall be as good as possible. In the third step, the set of solutions from step two are considered and so on. The result can be termed the *lexicographic minmax core*. This idea is somewhat similar to the nucleolus (see Sect. 2.3.5). However, the nucleolus measures the so-called excess $c(S) - \sum_{i \in S} \pi_i$ (i.e., benefit) in absolute terms while the concept of the minmax core uses η as a percentage; i.e., relative measure.

As for the core and the nucleolus, the concept of the proportional least core is enhanced to the notion of the *proportional nucleolus* by Young et al. (1982) and Lemaire (1984) which can be compared to the above described lexicographic minmax core. Since the work of Faigle et al. (1998b), this concept is also known as the *nucleon*. The excess for the nucleon is defined by

$$e(S, \pi) = \frac{c(S) - \sum_{i \in S} \pi_i}{c(S)}.$$

Lemaire (1984) describes the difference between the nucleolus and the proportional nucleolus as follows: "In one case coalitions are taxed in order to make the core exist, in the other case coalitions are subsidized in order to reduce the core to a single imputation."

The *disruptive nucleolus* uses the following excess function (Littlechild and Vaidya 1976):

$$e^d(S, \pi) = \frac{c(N \setminus S) - \sum_{i \in N \setminus S} \pi_i}{c(S) - \sum_{i \in S} \pi_i}.$$

The propensity to disrupt for a coalition S is defined as the ratio between the amount $N \setminus S$ and S would lose if the allocation according to the disruptive nucleolus is abandoned. By construction, this variant also belongs to the core if the core is not empty.

2.3.7 The Subcoalition-Perfect Core

Assume that the characteristic function c is not monotone – however, it may be subadditive. In such a case, there exists a coalition S_1 and supercoalition $S_2 \supset S_1$ ($S_2 \subseteq N$) such that $c(S_2) < c(S_1)$. This means for $S_2 \subset N$ that the grand coalition and an arbitrary core cost allocation $\pi \in C(N, c)$ may give S_1 an incentive to leave the grand coalition even though c is subadditive; i.e., S_1 may prefer the game (S_2, c) over the game (N, c). This may direct S_1 to force $S_2 \setminus S_1$ to form a coalition S_2 instead of the grand coalition. Drechsel and Kimms (2010c) investigate this behavior and describe it as a "force of anarchy" in the game because the cost allocation $\pi \in C(N, c)$ may not be considered as stable any longer if $c(S_2) < \sum_{i \in S_1} \pi_i \leq c(S_1)$ is true.

Look at a small example for illustration: We have a subadditive but non-monotone cooperative cost game – the values for the characteristic function are displayed in Table 2.2. Obviously, due to $c(\{2, 3\}) < c(\{3\})$, the game is not monotone. Assume $\pi = (0, -2, 9)$, player 3 would prefer the game $(\{2, 3\}, c)$ over the game (N, c).

Table 2.2 Numerical example for the subcoalition-perfect core in a cost game

$c(1)$	$c(2)$	$c(3)$	$c(1,2)$	$c(1,3)$	$c(2,3)$	$c(1,2,3)$
4	5	9	8	10	7	9

Based on these findings, Drechsel and Kimms (2010c) develop a new core concept which they call the *subcoalition-perfect core* to assure core cost allocations where no coalition S_1 has to carry higher costs in the grand coalition than in any other supercoalition S_2 of S_1. Mathematically, this can be defined as

$$C^{SCP}(N,c) = \left\{ \pi \in \mathbb{R}^{|N|} \,\middle|\, \sum_{i \in N} \pi_i = c(N) \text{ and } \sum_{i \in S_1} \pi_i \le c(S_2) \right.$$
$$\left. \text{for all } S_1 \subset N, \ S_1 \ne \emptyset \text{ and } S_1 \subseteq S_2 \subseteq N \right\}.$$

An equivalent formulation is

$$C^{SCP}(N,c) = \left\{ \pi \in \mathbb{R}^{|N|} \,\middle|\, \sum_{i \in N} \pi_i = c(N) \text{ and } \sum_{i \in S_1} \pi_i \le \min_{S_1 \subseteq S_2 \subseteq N} c(S_2) \right.$$
$$\left. \text{for all } S_1 \subset N, \ S_1 \ne \emptyset \right\}.$$

Obviously, $C^{SCP}(N,c) \subseteq C(N,c)$ holds in general and $C^{SCP}(N,c) = C(N,c)$ is true if c is monotone. Furthermore, Drechsel and Kimms (2010c) provide a proof regarding an interesting property of the subcoalition-perfect core: The subcoalition-perfect core equals the set of non-negative core allocations

$$C^+(N,c) = \{\pi \in C(N,c) | \pi_i \ge 0 \text{ for all } i \in N\}.$$

The proof that $C^+(N,c) = C^{SCP}(N,c)$ can be described as follows (see Drechsel and Kimms 2010c): If the characteristic function is monotone, the proof can be derived immediately from the property described in Sect. 2.3.4 ($\pi_i \ge 0$ for all i if c is monotone) and from the fact that for all coalitions S_1

$$c(S_1) = \min_{S_1 \subseteq S_2 \subseteq N} c(S_2).$$

Hence, $C^+(N,c) = C(N,c) = C^{SCP}(N,c)$ if c is monotone. What happens if c is not monotone?

"$C^+(N,c) \subseteq C^{SCP}(N,c)$": In case of an empty set of non-negative core allocations, the proof would be trivial. Thus, assume that the set of non-negative core allocations is nonempty. Assume furthermore that there exists a non-negative core allocation $\pi \in C^+(N,c)$ that is not in the subcoalition-perfect core. Which means that two coalitions $S_1 \subseteq S_2 \subseteq N$ exist with $c(S_2) < \sum_{i \in S_1} \pi_i \leq c(S_1)$. The chosen core allocation π fulfills

$$\sum_{i \in S_1} \pi_i + \sum_{i \in S_2 \setminus S_1} \pi_i = \sum_{i \in S_2} \pi_i \leq c(S_2) < \sum_{i \in S_1} \pi_i.$$

For this reason, there must be at least one player i in $S_2 \setminus S_1$ with a negative cost share ($\pi_i < 0$). Mathematically, this can be expressed by the following implication:

$$\exists S_1 \subset N : \exists S_1 \subset S_2 \subset N : \sum_{i \in S_1} \pi_i > c(S_2) \Rightarrow \exists i \in N : \pi_i < 0.$$

This contradicts to the before stated assumption $\pi \in C^+(N, c)$ and by contraposition we get

$$\forall i \in N : \pi_i \geq 0 \Rightarrow \forall S_1 \subset N : \forall S_1 \subset S_2 \subset N : \sum_{i \in S_1} \pi_i \leq c(S_2).$$

Hence, $\pi \in C^{SCP}(N, c)$ must hold.

"$C^+(N, c) \supseteq C^{SCP}(N, c)$": The proof would be trivial if the set of subcoalition-perfect core allocations is empty. Therefore, let this set be nonempty. Assume that there exists a subcoalition-perfect core allocation $\pi \in C^{SCP}(N, c)$ with $\pi_i < 0$ for at least one player $i \in N$; i.e., $\pi \notin C^+(N, c)$. The chosen cost allocation is efficient and, therefore, we have $c(N) = \sum_{j \in N} \pi_j$. Due to $\pi_i < 0$, we get $\sum_{j \in N} \pi_j < \sum_{j \in N \setminus \{i\}} \pi_j$. But π is an element form the subcoalition-perfect core and therefore $\sum_{j \in N \setminus \{i\}} \pi_j \leq c(N)$ must hold which leads to the conclusion that

$$c(N) = \sum_{j \in N} \pi_j < \sum_{j \in N \setminus \{i\}} \pi_j \leq c(N).$$

Hence, $\pi \in C^+(N, c)$ must be true. Summarizing the proof, $C^+(N, c) = C^{SCP}(N, c)$ holds. In Chap. 7, we will apply the subcoalition-perfect core to a practical non-monotone game.

The concept of the subcoalition-perfect core can be applied to *profit games* as well. Assume a profit game (N, v) where the characteristic function v is not monotone; i.e., there might be a coalition S_1 and a supercoalition $S_2 \supset S_1$ ($S_2 \subseteq N$) such that $v(S_2) < v(S_1)$. An allocation π would not be preferable if $v(S_2) \leq \sum_{i \in S_2} \pi_i < v(S_1) \leq \sum_{i \in S_1} \pi_i$ ($S_1 \subseteq S_2 \subseteq N$). Thus, the subcoalition-perfect core in case of profit games contains allocations where no coalition S_2 has a lower payoff in the grand coalition than in any subcoalition S_1 of S_2.

$$C^{SCP}(N, v) = \left\{ \pi \in \mathbb{R}^{|N|} \,\middle|\, \sum_{i \in N} \pi_i = v(N) \text{ and } \sum_{i \in S_2} \pi_i \geq v(S_1) \right.$$

$$\left. \text{for all } S_1 \subset N, \ S_1 \neq \emptyset \text{ and } S_1 \subseteq S_2 \subseteq N \right\}.$$

The equivalent formulation is

$$C^{SCP}(N,v) = \left\{ \pi \in \mathbb{R}^{|N|} \,\middle|\, \sum_{i \in N} \pi_i = v(N) \text{ and } \sum_{i \in S_2} \pi_i \geq \max_{S_1 \subseteq S_2 \subseteq N} v(S_1) \right.$$
$$\left. \text{for all } S_1 \subset N, \; S_1 \neq \emptyset \right\}.$$

Again, $C^{SCP}(N,v) \subseteq C(N,v)$ and if v is monotone, $C^{SCP}(N,v) = C(N,v)$ holds. Profit games behave differently in the monotonicity proof concerning positive core allocations (see p. 24):

$$\pi_i = v(N) - \sum_{j \in N \setminus \{i\}} \pi_j \leq v(N) - v(N \setminus \{i\}) \geq 0 \quad \text{for all } i \in N.$$

Hence, there is no general proof that in monotone profit games the profit shares π_i are always non-negative. But we can prove that:
"$C^+(N,v) \subseteq C^{SCP}(N,v)$": Assume that there exist non-negative core allocations and a non-negative core allocation π which is not in the subcoalition-perfect core; i.e., $S_1 \subseteq S_2 \subseteq N$ with $v(S_2) \leq \sum_{i \in S_2} \pi_i < v(S_1)$. For the chosen core allocation, π must hold

$$v(S_2) \leq \sum_{i \in S_1} \pi_i + \sum_{i \in S_2 \setminus S_1} \pi_i = \sum_{i \in S_2} \pi_i < v(S_1) \leq \sum_{i \in S_1} \pi_i.$$

Hence, there has to be at least one player in $S_2 \setminus S_1$ with a negative profit share ($\pi_i < 0$).

$$\exists S_1 \subset N : \exists S_1 \subset S_2 \subset N : \sum_{i \in S_2} \pi_i < v(S_1) \Rightarrow \exists i \in N : \pi_i < 0.$$

Hence,

$$\forall i \in N : \pi_i \geq 0 \Rightarrow \forall S_1 \subset N : \forall S_1 \subset S_2 \subset N : \sum_{i \in S_2} \pi_i \geq v(S_1).$$

Accordingly, $\pi \in C^{SCP}(N,v)$ holds.

Contrary to cost games and according to the relation between monotonicity and non-negative profit shares, $C^+(N,v) \supseteq C^{SCP}(N,v)$ does not hold: Assume that there exists an allocation in the subcoalition-perfect core and that there is an allocation in the subcoalition-perfect core with at least one $\pi_i < 0$. The efficiency constraint $v(N) = \sum_{j \in N} \pi_j$ holds. Due to $\pi_i < 0$, we have

$$v(N) = \sum_{j \in N} \pi_j < \sum_{j \in N \setminus \{i\}} \pi_j \geq v(N \setminus \{i\}).$$

Table 2.3 Numerical example for the subcoalition-perfect core in a profit game

$v(1)$	$v(2)$	$v(3)$	$v(1,2)$	$v(1,3)$	$v(2,3)$	$v(1,2,3)$
-1	0	-1	2	-1.5	2	4

For an allocation in the subcoalition-perfect core must hold $\sum_{j \in N} \pi_j \geq v(N \setminus \{i\})$. Therefore, we might have negative cost allocations in the subcoalition-perfect core which can be supported by a small numerical example. Table 2.3 displays the characteristic function for a profit game with three players. The game is superadditive, but not monotone (e.g., $v(1) > v(1,3)$). The following constraints define the core and the subcoalition-perfect core for this game:

$$
\begin{array}{ll}
C(N,v): & \pi_1 + \pi_2 + \pi_3 = 4 \\
 & \pi_1 \geq -1 \\
 & \pi_2 \geq 0 \\
 & \pi_3 \geq -1 \\
 & \pi_1 + \pi_2 \geq 2 \\
 & \pi_1 + \pi_3 \geq -2 \quad \leftarrow \\
 & \pi_2 + \pi_3 \geq 2
\end{array}
\qquad
\begin{array}{ll}
C^{SCP}(N,v): & \pi_1 + \pi_2 + \pi_3 = 4 \\
 & \pi_1 \geq -1 \\
 & \pi_2 \geq 0 \\
 & \pi_3 \geq -1 \\
 & \pi_1 + \pi_2 \geq 2 \\
 & \pi_1 + \pi_3 \geq -1 \quad \leftarrow \\
 & \pi_2 + \pi_3 \geq 2
\end{array}
$$

An allocation in the subcoalition-perfect core is, e.g., $\pi_1 = -1$, $\pi_2 = 5$, $\pi_3 = 0$.

Thus, we can summarize that the subcoalition-perfect core for a cost game equals the set of non-negative core allocations, but for profit games, the set of non-negative core allocations is only a subset of the subcoalition-perfect core.

2.3.8 The Interval Core

In Sect. 2.2.4, we introduced interval-valued games that can handle uncertainties of the characteristic function. To deal with practical interval-valued games, we need to define an allocation concept that can handle interval-valued characteristic functions.

We use the following notation for preparing a core variant that applies to interval-valued games. The sum of two intervals $I = [\underline{I}; \overline{I}]$ and $J = [\underline{J}; \overline{J}]$ is defined to be $I + J = [\underline{I} + \underline{J}; \overline{I} + \overline{J}]$. The interval I is called weakly better than interval J ($I \preceq J$ or $J \succeq I$), if and only if $\underline{I} \leq \underline{J}$ and $\overline{I} \leq \overline{J}$. Following for instance Alparslan-Gök et al. (2008a), the *interval core* of a cooperative interval-valued game can be defined as follows:

$$
C^{IV}(N, c^{IV}) = \left\{ (I_1, \ldots, I_{|N|}) \in I(\mathbb{R})^{|N|} \,\middle|\, \sum_{i \in N} I_i = c^{IV}(N) \text{ and } \sum_{i \in S} I_i \preceq c^{IV}(S) \right.
$$
$$
\left. \text{for all } S \subset N, \ S \neq \emptyset \right\}.
$$

Hence, the interval core consists of vectors that aggregate the interval shares for every player in the grand coalition. These interval shares have to fulfill the known efficiency and stability constraints.

2.3.9 The Shapley Value

Apart from the core, its manifold variants, and related concepts, a lot of other solution concepts in cooperative game theory exist. In this section, we will shortly introduce the most common of them for the sake of completeness and will refer to more substantial contributions for a further discussion. Fromen (2004) gives an extensive survey of the existing solution concepts including evaluations and comparisons (see Fromen 2004, in particular Chap. 4).

Shapley (1953) aims to develop an allocation method that yields a unique solution for every game in coalitional form and follows three axioms. The axioms are derived from properties that should be satisfied by such an allocation. Before describing the axioms, we have to define a *permutation* $P : N \rightarrow N$ (a one to one mapping onto itself) of the set of players N such that, for every $j \in N$, there exists exactly one $i \in N$ that $P(i) = j$. Furthermore, we let Pc be the coalitional game such that

$$Pc(\{P(i) | i \in S\}) = c(S) \quad \text{for all } S \subseteq N.$$

That means, player i in c plays essentially the same role as player $P(i)$ in Pc. A coalition R is a *carrier* of a game if and only if

$$c(S \cap R) = c(S) \quad \text{for all } S \subseteq N.$$

Accordingly, all players $S \backslash R$ are called *dummies* in the game because their presence in the game does not change its value. The value Φ denotes a function which associates a real number with each $i \in N$ and satisfies the following axioms (see Shapley 1953):

- Symmetry: for any permutation and any player i, $P \in P(N)$, $\Phi_{P(i)}(Pc) = \Phi_i(c)$
- Efficiency: for any carrier R, $\sum_{i \in R} \Phi_i = c(R)$
- Additivity: for any two games (N, c_1) and (N, c_2), $\Phi(c_1 + c_2) = \Phi(c_1) + \Phi(c_2)$

In other words, the *Shapley value* yields an allocation where players are handled symmetrically. A cost share is assigned to players only if they actually generate costs. Furthermore, while combining two independent games, the players' values must be added player by player. Shapley (1953) proves that no further condition is needed to define a unique value. Additionally, he shows that some more properties hold assuming the three presented axioms and derives the following formula to determine the Shapley value:

$$\Phi_i(c) = \sum_{S \subseteq N \setminus \{i\}} \frac{|S|!(|N| - |S| - 1)!}{|N|!} [c(S \cup \{i\}) - c(S)]. \tag{2.9}$$

Myerson (1991) gives a very clear description regarding (2.9): Imagine the players randomly lining up in a queue at the door of a big room where they should all be assembled. There exist $|N|!$ different orders for the queue. For any set S not containing i, there are $|S|!(|N| - |S| - 1)!$ permutations so that S is the set of players standing ahead of i in the queue. Hence, the relation $|S|!(|N| - |S| - 1)!/|N|!$ defines the probability when i enters the room and S is already in there. In this case, player i's marginal contribution is $c(S \cup \{i\}) - c(S)$. Thus, the Shapley value can be interpreted as the expected marginal contribution of every player i. Until today, the Shapley value is very famous in business and mathematical literature (see e.g., Roth 1988).

The Shapley value might not be inside the core (see Lemaire 1984, p. 72, for a numerical example), but Shapley (1971) prove that in case of a convex profit game, the Shapley value always belongs to the core – it displays the center of gravity of the core's extremal points. For instance, Hamlen et al. (1980, p. 270) explain that the uniqueness of the Shapley value may not be an advantage in any case because it does not permit any flexibility (e.g., to achieve other goals set by management). They present a generalization of the Shapley allocation to restore this problem.

2.3.10 Conclusions

From the viewpoint of practical applications, the first target of cost allocation should be to ensure stability of the cooperation. No cooperation will exist a longer period of time if any player thinks that it could do better without the other participants in the cooperation. Secondly, a cost allocation can offer incentives such that the cooperating partners act in the sense of the grand coalition. Lemaire (1984, p. 63) proves with a small example that allocations where one partner is subsidized at the expense of another may not assure optimal total costs for the grand coalition. To assure stability, we introduce the so-called coalitional rationality property; i.e., no player or subcoalition should be better off when defecting from the grand coalition.

A core cost allocation guarantees these properties and displays a very intuitive and easy to comprehend concept. However, the concept of the core has two main drawbacks: The core may be empty and if it exists, it is rarely unique. The first dis-advantage can be solved by using core variants like the least core or the minmax core where it is possible to violate stability constraints to a certain amount (absolute or relative). The second disadvantage can be seen as a chance to integrate other desir-able properties or providing flexibility for management decisions. Neither allocation methods from accounting nor the Shapley value can assure stable cost allocations for general cooperative games.

The downside of cost allocation methods coming from cooperative game theory is that they require much more information than classical accounting approaches.

Normally, due to the coalitional rationality, we need information about the characteristic function values $c(S)$ for all subcoalitions $S \subseteq N$; i.e., data for $2^{|N|} - 1$ cost values are necessary. The application in many practical situations is therefore limited.

In the following chapter, we will tie in with these findings and present a procedure which allows to compute stable cost allocations without utilizing all characteristic function values. For more extensive discussions regarding the suitability of the different cooperative game theory approaches, compare, e.g., Young et al. (1982), Lemaire (1984), Fromen (2004), and Zelewski (2007).

Chapter 3
Algorithmic Game Theory

Most of the methods developed in the field of game theory are more or less conceptual tools that should predict rational strategic behavior of individuals in conflicting or cooperating situations. But an equilibrium concept or an allocation method would lose much of its credibility if it is not efficiently computable. Algorithmic game theory tries to resolve this lack.

Nisan et al. (2007) describe *algorithmic game theory* as being at the interface of computer science, game theory, and economic theory. Thus, algorithmic game theory combines algorithmic thinking with game theoretic and/or economic concepts. During the last years, algorithmic game theory has been an emerging field of research. The first contributions regarding algorithmic game theory study the most prominent tool in game theory: the Nash equilibrium (see Papadimitriou 2007). For further contributions of algorithmic game theory, we refer to the substantial textbook Nisan et al. (2007). Coming back to the settings of cooperative game theory described in the previous chapter, we will now concentrate our explanations on cooperative algorithmic game theory.

One possibility to provide tools, making game theory methods computable, is to utilize operations research knowledge. The interrelation between operations research and non-cooperative game theory is well established. However, contributions regarding operations research and cooperative game theory are of a more recent date (see Borm et al. 2001, p. 140). Apart from that, operations research offers many decision problems where cooperative game theory methods are applicable. Borm et al. (2001) coined the term *operations research games* for games with an interplay between optimization and allocation. Due to the following applications (see Chap. 5 and the following), the present work can be sorted into this field of research as well.

Section 3.1 gives a short survey of contributions that belong to the field of algorithmic game theory and that are closely related to cooperative settings, our preferred allocation method the core, and linear programming. Afterwards, we develop an algorithm for efficient computing of core allocations and discuss its application for core variants.

J. Drechsel, *Cooperative Lot Sizing Games in Supply Chains*, Lecture Notes
in Economics and Mathematical Systems 644, DOI 10.1007/978-3-642-13725-9_3,
© Springer-Verlag Berlin Heidelberg 2010

3.1 Literature

There are several contributions that derive *complexity results* regarding varying cooperative games: Deng and Papadimitriou (1994) study the complexity of different solution concepts – the Shapley value, the core, the kernel, the nucleolus, the ϵ-core, and the bargaining set. Faigle et al. (1997) prove that testing core membership of an imputation for the minimum cost spanning tree game is $\mathcal{NP}$-complete. The related fixed cost spanning forest problem is investigated by Granot and Granot (1992). They prove that core elements can be determined in strongly polynomial time. Fang et al. (2002) extend this proposition to flow games and linear production games. Several different games on graphs are considered by Deng et al. (1999) who derive similar results concerning complexity. Deng et al. (2000) study the total balancedness of several combinatorial optimization games, e.g., partition games, packing and covering games. Goemans and Skutella (2004) investigate facility location games and verify that testing the core membership for such games is $\mathcal{NP}$-complete. Additionally, they show that for a nonempty core, this problem can be solved in polynomial time as well as computing a core element.

Several papers use *linear programming* with the core: Owen (1975) introduces linear production games where a point in the core can be obtained by solving a linear program. Using the dual variables of a linear programming formulation of a problem for defining a core element is an important contribution because this can be successfully adopted in other applications. For instance, Chen and Zhang (2006) use duality theory to find a core allocation for a special economic lot sizing game (it is defined differently from the ELS game following in Chap. 5). Granot (1986) studies linear production games as well, but generalizes the assumption concerning additivity of resources. Apart from their complexity results, Deng et al. (1999) observe that the core for their games on graphs is nonempty if and only if an associated linear program has an integer optimal solution.

Problem specific algorithms for computing core allocations has been developed in the literature: Derks and Kuipers (1997) introduce routing games and investigate the non-emptiness of the core depending on the selected tour. Furthermore, they present an $\mathcal{O}(n^2)$ algorithm to compute a core allocation if the core is nonempty (n is the number of customers). Information graph games (a subclass of minimum cost spanning tree games) are studied by Kuipers (1993). He derives that in a graph game with n players the core can be described by a set of at most $(2n - 1)$ linear constraints. Sánchez-Soriano (2006) shows for the transportation game that every core element is contained in a pairwise solution using a specific weight system. For the assignment game, Sotomayor (2003) discovers that if an instance has only one optimal matching, the core has infinitely many payoffs, whereas instances with a unitary core have more than one optimal matching. Borm et al. (2001) give an extensive survey of examples regarding procedures to compute core elements. They particularly address operations research games that are organized into five categories: connection (fixed tree, spanning tree), routing (Chinese postman, traveling salesman), scheduling (sequencing, permutation, assignment), production (linear production, network flow), and inventory. In Chap. 5, we will study inventory games in more detail.

Apart from those tailor-made algorithms for specific games, there are contributions concerning *approximations and core related concepts*. Kamiya and Talman (1991) present an algorithm for balanced games where the subdivision of an appropriate simplex into smaller simplices yields a simplex that provides an approximating core element. Hallefjord et al. (1995) develop a constraint generation approach for computing the nucleolus in linear production games. This idea of a constraint generation approach will be applied in the next section. Göthe-Lundgren et al. (1996) also use row generation to compute the nucleolus but for a vehicle routing game. In this case, the subproblem is a hard-to-solve mixed-integer programming problem. Chardaire (2001) presents comments on the paper of Göthe-Lundgren et al. (1996). Apart from the row generation approaches, Faigle et al. (2001) provide an algorithm to compute the nucleolus that is based on the ellipsoid method and Maschler's scheme for approximating the prekernel.

3.2 Computing Core Cost Allocations

As introduced in Sect. 2.3.4 ((2.7), p. 22), the core specifies a constraint satisfaction problem. According to the stability restrictions, the number of core constraints is exponential with an order of magnitude equal to $2^{|N|}$. That is to say, the number of coalitions rises exponentially with an increasing number of players. Due to this phenomenon, it might be helpful to use an algorithm that finds an element in the core without testing the constraints for all possible coalitions S. Drechsel and Kimms (2010a) suggest a row generation procedure to tackle the problem. Basically, we start with a relaxed version of the original problem and add missing relaxed constraints over several iterations. The target is to get an optimal solution without adding all restrictions.

The master problem of the procedure is based on the constraint satisfaction problem given by the core definition (2.7) and can be formulated as the following linear program:
$MP(\mathcal{S})$:

$$\min w \tag{3.1}$$

s.t.

$$\sum_{i \in N} \pi_i = c(N) \tag{3.2}$$

$$\sum_{i \in S} \pi_i - w \le c(S) \qquad\qquad S \in \mathcal{S} \tag{3.3}$$

$$\pi_i \in \mathbb{R} \qquad\qquad i \in N \tag{3.4}$$

$$w \ge 0. \tag{3.5}$$

The decision variables π_i give the cost share for every player i. The values $c(N)$ and $c(S)$ ($S \in \mathcal{S}$) are parameters; i.e., they have to be computed in advance. This computational burden is inherent to the definition of the core which assumes that all values $c(S)$ for $S \subseteq N$, $S \neq \emptyset$, are known in advance. Let $\mathcal{S}$ be a set of coalitions $S \subset N$ for which the stability condition in the core explicitly be stated. That means, if we choose $\mathcal{S} = 2^N \setminus \{\emptyset, N\}$, we would directly compute a core element. The decision variable w is used to detect whether or not the core is empty. If the optimal objective function value yields $w = 0$, then the core for this instance is nonempty and the corresponding π_i-values define a cost allocation in the core. In contrast, if the optimal solution yields $w > 0$, we cannot find an element in the core for the examined instance; i.e., the core is empty.

After solving the master problem, we get a cost allocation $(\pi_1, \pi_2, \ldots, \pi_{|N|})$. Obviously, the model is controlling only some of the subcoalitions (see (3.3)). Hence, it is now required to find an arbitrary coalition $S' \subset N$ ($S' \notin \mathcal{S}$) for which this allocation is not in the core:

$$\sum_{i \in S'} \pi_i > c(S'). \tag{3.6}$$

If there exists such a coalition S', the corresponding minimum total costs $c(S')$ have to be computed and a new constraint will be added to the master problem:

$$\sum_{i \in S'} \pi_i - w \leq c(S'). \tag{3.7}$$

The master problem is solved again and gives a new cost allocation $(\pi_1, \pi_2, \ldots, \pi_{|N|})$. We can test this new allocation again whether there is still any coalition S' not holding the core restrictions. The algorithm stops when there is no coalition S' anymore violating the core constraints. Summarized, the algorithm proceeds in the following steps:

1. Define a small initial set $\mathcal{S}$; e.g., $\mathcal{S} = \{\{1\}, \{2\}, \ldots, \{|N|\}\}$. Compute the individual total costs $c(S)$ for those coalitions $S \in \mathcal{S}$ and the total costs $c(N)$ for the coalition N.
2. Solve the master problem $MP(\mathcal{S})$ (3.1)–(3.5) optimally.
3. If $w > 0$, stop the algorithm because the instance has an empty core.
4. Otherwise, find a coalition $S' \notin \mathcal{S}$ ($S' \neq \emptyset$) such that restriction (3.6) holds.
5. If no such coalition S' can be found, then stop the algorithm because the found allocation is in the core.
6. Otherwise, compute the total costs $c(S')$ for this coalition, add a constraint of type (3.7) to the master problem (update $\mathcal{S} = \mathcal{S} \cup \{S'\}$), and go to Step 2.

3.3 Theoretical Background

The idea of row generation for mathematical programming problems that have too many constraints in the original formulation is well established in the literature and has been already used for numerous applications. However, depending on the concrete application, the algorithm differs. This section shall give an overview of the theoretical background of row generation and a short survey about successful applications of row generation procedures.

Sethi and Thompson (1984) introduce a *pivot and probe algorithm* to refine the simplex algorithm. They prove that only 15–25% of the constraints of a linear program are candidates and 70–90% of those candidate constraints are tight at the optimum. With this knowledge, they develop the following algorithm: Starting with a relaxed version of the original linear program containing only candidate constraints, a lower bound is calculated. After that, a probe in the form of a line segment is used to find a most violated constraint not contained in the relaxed problem. This constraint is added to the relaxed problem and it is solved again to get a new and hopefully better upper bound. In Thompson and Sethi (1986), an application of the pivot and probe algorithm to generalized transportation problems is presented.

Cutting plane procedures are related to our row generation procedure as well. This is firstly proposed by Gomory (1958). Cutting plane approaches has been developed to solve mixed integer linear programs. These concepts start with solving an LP relaxation of the original problem and try to find additional inequalities that cut the optimal solution of the LP relaxation but no integral solution for the original problem. See e.g., Schrijver (1986, pp. 339 and 354) and Bradley et al. (1977, p. 402) for a detailed description.

Benders decomposition (primal decomposition) can be seen as another relative of our row generation procedure. Benders (1962) introduces a decomposition approach which is a kind of resource directive decomposition that alternately solves a mixed integer linear program with only one non-integral variable and a linear program. At each iteration, either an optimal solution is found or a new violated inequality, see, e.g., Schrijver (1986, p. 371). A decade later, Geoffrion (1972) proposes an extension that makes Benders decomposition applicable to situations where the subproblem needs not be a linear program as well. See Florian et al. (1976) and Geoffrion and Graves (1974) for early applications of Benders decomposition to mixed integer programs. Magnanti and Wong (1981) present an acceleration technique to reduce the number of iterations of the Benders decomposition and discuss "proper" formulations of mixed integer programs to generate stronger cuts for the decomposition.

Most closely related to our approach is the contribution of Hallefjord et al. (1995). They compute the nucleolus by means of row generation based on the idea of Gilmore and Gomory (1961). The underlying cooperative game is based on a simple linear program. In Bonnans and André (2008), a row generation procedure to compute the least core and the prenucleolus for linear and convex production games is presented where the most violated constraint is found via lower estimates of the cost function. Göthe-Lundgren et al. (1996) present a cooperative game in the field of vehicle routing (several players can cooperate to consolidate transportation tasks in joint tours). In this paper, the nucleolus is computed by row generation as well, but now for a hard-to-solve mixed-integer programming problem (the VRP). See Chardaire (2001) for a note on the work of Göthe-Lundgren et al. (1996). In contrast to our approach, the number of constraints considered in the subproblems strictly increases with every iteration (which is due to the definition of the nucleolus) while the size of our subproblem remains the same throughout (compare Sect. 5.2.1). Another recent application of row generation to compute the nucleolus can be found in Çetiner and Kimms (2009) where cost allocations for strategic airline alliances are computed.

Apart from other approaches in the literature, duality theory might help to understand the theoretical background of our procedure. Consider the dual formulation of the introduced master problem $MP(\mathcal{S})$:

$DMP(\mathcal{S})$:

$$\max \lambda c(N) - \sum_{S \in \mathcal{S}} \mu_S c(S)$$

s.t.

$$\sum_{S \in \mathcal{S}} \mu_S \leq 1$$

$$\lambda - \sum_{S \in \mathcal{S}: i \in S} \mu_S = 0 \qquad\qquad i \in N$$

$$\lambda \in \mathbb{R}$$

$$\mu_S \geq 0 \qquad\qquad S \in \mathcal{S}.$$

The dual variable λ corresponds to constraint (3.2) and μ_S to (3.3). Solving $DMP(\mathcal{S})$ by means of column generation is equivalent to our proposed row generation procedure. Column generation is used for linear programs with a huge number of variables – the number of variables is reduced at the beginning and with the help of a subproblem, new variables will be integrated in the starting problem successively as long as the objective function value of the initial problem can be improved. At first, Dantzig and Wolfe (1960) theoretically discuss a decomposition principle for solving linear programs more easily which is later on called column generation. See also standard textbooks and surveys; e.g., Bradley et al. (1977, p. 540) and Soumis (1997), for a further general discussion of column generation procedures and their manifold applications. Barnhart et al. (1998) discuss column generation methods as well. Due to the auspicious results reached with column generation procedures, our approach with row generation seems to be promising as well.

3.4 Including Fairness Criteria

The before presented row generation algorithm answers the question whether or not the core of the game is empty. If the core is not empty, the procedure computes a core allocation. Due to the fact that the core does not consist of a single element, in general, the following question arises: Which core element should be chosen? Hence, we extend the formulation of the master problem to find a core cost allocation with a certain characteristic. Obviously, this only makes sense for instances with a nonempty core which can be verified by running the before presented algorithm. Apart from that, every core element is stable, an arbitrary core element might not be seen as fair although nobody has an incentive to leave the grand coalition because some players may profit more than others from the cooperation. Drechsel and Kimms (2010a) introduce several reformulations of the master problem to overcome this problem:

Variant 1: The absolute cost shares should deviate as little as possible among the players.

$MP^I(\mathcal{S})$:

$$\min \overline{\Pi} - \underline{\Pi}$$

s.t.

$$\sum_{i \in N} \pi_i = c(N)$$

$$\sum_{i \in S} \pi_i \leq c(S) \qquad\qquad S \in \mathcal{S}$$

$$\overline{\Pi} \geq \pi_i \qquad\qquad i \in N$$

$$\underline{\Pi} \leq \pi_i \qquad\qquad i \in N$$

$$\pi_i \in \mathbb{R} \qquad\qquad i \in N$$

$$\overline{\Pi}, \underline{\Pi} \in \mathbb{R}.$$

The values $\overline{\Pi}$ and $\underline{\Pi}$ denote the highest and lowest cost share over all players, respectively. In combination with the objective function, the difference between those values should be minimized. Drechsel and Kimms (2010a) note that this master problem formulation yields an equal distribution of the cost shares; i.e.,

$$\pi_i = \frac{c(N)}{|N|},$$

if and only if

$$\frac{|S|}{|N|} \geq \frac{c(S)}{c(N)} \text{ for all } S \subseteq N$$

holds. This is easy to prove for an arbitrary coalition S:

$$c(S) \geq \sum_{i \in S} \pi_i = \sum_{i \in S} \frac{c(N)}{|N|} = \frac{|S|}{|N|} c(N).$$

Geometrically, such a core element is closest according to the Euclidean metric to the midpoint of the polyhedron $\{\pi \geq 0| \sum_{i \in N} \pi_i = c(N)\}$. The midpoint of the polyhedron represents the equal distribution of the cost shares.

Variant 2: The percentage cost savings should deviate as little as possible among the players.

$MP^{II}(\mathcal{S})$:

$$\min \overline{\Pi} - \underline{\Pi}$$

s.t.

$$\sum_{i \in N} \pi_i = c(N)$$

$$\sum_{i \in S} \pi_i \leq c(S) \qquad\qquad S \in \mathcal{S}$$

$$\overline{\Pi} \geq \pi_i / c(\{i\}) \qquad\qquad i \in N$$

$$\underline{\Pi} \leq \pi_i / c(\{i\}) \qquad\qquad i \in N$$

$$\pi_i \in \mathbb{R} \qquad\qquad i \in N$$

$$\overline{\Pi}, \underline{\Pi} \in \mathbb{R}.$$

It is shown by Drechsel and Kimms (2010a) that a distribution of the cost shares proportional to the $c(\{i\})$ values; i.e.,

$$\pi_i = \frac{c(\{i\})}{\sum_{j \in N} c(\{j\})} c(N),$$

defines a core element if and only if

$$\frac{\sum_{i \in S} c(\{i\})}{\sum_{i \in N} c(\{i\})} \leq \frac{c(S)}{c(N)} \text{ for all } S \subseteq N$$

is true. This can be proved for an arbitrary coalition S:

$$c(S) \geq \sum_{i \in S} \pi_i = \sum_{i \in S} \frac{c(\{i\})}{\sum_{j \in N} c(\{j\})} c(N) = \frac{\sum_{i \in S} c(\{i\})}{\sum_{j \in N} c(\{j\})} c(N).$$

Interpreting the result of $MP^{II}(\mathcal{S})$ from a geometric perspective, such a core element is closest according to the Euclidean metric to a specific convex combination of the $|N|$ vertices of the polyhedron $\{\pi \geq 0| \sum_{i \in N} \pi_i = c(N)$ and $\pi_i \leq c(\{i\})$ for all $i \in N\}$ where $\omega_i = c(\{i\})/\sum_{j \in N} c(\{j\})$ is the weight of the vertex κ_i with $\pi_i \leq c(\{i\})$ being unnecessary to define that particular vertex κ_i. We will give a numerical example in Sect. 5.2.2 to illustrate these statements.

Other kinds of fairness aspects are also conceivable, for instance, one where the difference between individual costs and cost shares of each player should be preferably equally distributed among the players:

$$\min \overline{\Pi} - \underline{\Pi}$$

s.t.

$$\sum_{i \in N} \pi_i = c(N)$$

$$\sum_{i \in S} \pi_i \leq c(S) \qquad\qquad S \in \mathcal{S}$$

$$\overline{\Pi} \geq c(\{i\}) - \pi_i \qquad\qquad i \in N$$

$$\underline{\Pi} \leq c(\{i\}) - \pi_i \qquad\qquad i \in N$$

$$\pi_i \in \mathbb{R} \qquad\qquad i \in N$$

$$\overline{\Pi}, \underline{\Pi} \in \mathbb{R}.$$

In order to stretch this topic not too excessively, we will study only the first two presented variants in Chap. 5.

3.5 Computing Core Variants

In Sect. 2.3, we presented several variants of the core. We now want to show that the proposed row generation procedure is applicable for all of these variants with only slight changes to be made.

For the ϵ-*core*, the master problem needs adjustment in the stability constraint (see Drechsel and Kimms 2010a).

$MP^\epsilon(\mathcal{S})$:

$$\min w$$

s.t.

$$\sum_{i \in N} \pi_i = c(N)$$

$$\sum_{i \in S} \pi_i - w \leq c(S) + \epsilon \qquad\qquad S \in \mathcal{S}$$

$$\pi_i \in \mathbb{R} \qquad\qquad i \in N$$

$$w \geq 0.$$

Recall that ϵ is a parameter while computing the ϵ-core. The row generation algorithm proceeds as described on p. 44. ϵ will be a decision variable when computing a cost allocation in the *least core*. The variable w can be skipped and the objective function will now minimize ϵ (see Drechsel and Kimms 2010a).

$MP^L(\mathcal{S})$:

$$\min \epsilon$$

s.t.

$$\sum_{i \in N} \pi_i = c(N)$$

$$\sum_{i \in S} \pi_i \leq c(S) + \epsilon \qquad\qquad S \in \mathcal{S}$$

$$\pi_i \in \mathbb{R} \qquad\qquad i \in N$$

$$\epsilon \in \mathbb{R}.$$

We neglect Step 3 from the procedure, see p. 44, while computing an element in the least core because the least core is never empty.

The formulation for computing an element in the *minmax core* needs a small change in the stability constraint of the master problem (see Drechsel and Kimms 2010b).

$MP^M(\mathcal{S})$:

$$\min \eta \qquad\qquad\qquad (3.8)$$

s.t.

$$\sum_{i \in N} \pi_i = c(N) \qquad\qquad\qquad (3.9)$$

$$\sum_{i \in S} \pi_i \leq \eta\, c(S) \qquad\qquad S \in \mathcal{S} \qquad (3.10)$$

$$\pi_i \in \mathbb{R} \qquad\qquad i \in N \qquad (3.11)$$

$$\eta \in \mathbb{R}. \qquad\qquad\qquad (3.12)$$

Note that we can easily define $\eta \geq 0$: An optimal η will never be negative because comparing the stability constraints, it is obvious that a negative η forces negative π_i-values which contradicts the efficiency constraint as long as $c(N)$ is non-negative. As for the least core, we skip Step 3 in the row generation procedure (see p. 44) to calculate an element in the minmax core. In Sect. 2.3.6, we described the lexico-graphic minmax core. It is straightforward to extend the row generation procedure to compute an element in the lexicographic minmax core.

It was proved that the *subcoalition-perfect core* equals the set of non-negative core allocations (see Sect. 2.3.7) when dealing with cost games. Hence, we only have to define the decision variables π_i to be non-negative compared to the *MP*-formulation (3.1)–(3.5) (see Drechsel and Kimms 2010c).

$MP^+(\mathcal{S})$:

$$\min w \qquad\qquad\qquad (3.13)$$

s.t.

$$\sum_{i \in N} \pi_i = c(N) \tag{3.14}$$

$$\sum_{i \in S} \pi_i - w \le c(S) \qquad\qquad S \in \mathcal{S} \tag{3.15}$$

$$\pi_i \ge 0 \qquad\qquad i \in N \tag{3.16}$$

$$w \ge 0. \tag{3.17}$$

The row generation algorithm proceeds like described for the classical core (see p. 44). It is suggested to apply the subcoalition-perfect core to non-monotone games. In Chap. 7, we will present the capacitated lot sizing game which is not monotone in general. A computational study will show the application of the minmax core and the subcoalition-perfect core.

The procedure regarding the computation of an element in the *approximate core* is similar. Computing the nucleolus, however, requires some extension: A sequence of nested optimization problems has to be solved to obtain a lexicographic optimum (see e.g., Hallefjord et al. 1995; Faigle et al. 2001; Çetiner and Kimms 2009).

3.6 Computing Interval Core Elements

After defining the interval core (see Sect. 2.3.8), we now want to extend the row generation procedure regarding interval-valued games to compute an element in the interval core.

To use the mentioned procedure as a subroutine, we need to reformulate interval-valued games and the interval core. Following Drechsel and Kimms (2009), we specify an interval-valued game by a triple $(N, \underline{c}^{IV}, \overline{c}^{IV})$ (with the set of players N and $[\underline{c}^{IV}; \overline{c}^{IV}]$ as the closed interval for the characteristic function). Now, two sets of extreme cost allocations can be defined as

$$\underline{C}^{IV}(N, \underline{c}^{IV})$$

$$= \left\{ \pi \in \mathbb{R}^{|N|} \ \middle| \ \sum_{i \in N} \pi_i = \underline{c}^{IV}(N) \text{ and } \sum_{i \in S} \pi_i \le \underline{c}^{IV}(S) \text{ for all } S \subset N, \ S \ne \emptyset \right\} \tag{3.18}$$

and

$$\overline{C}^{IV}(N, \overline{c}^{IV})$$

$$= \left\{ \pi \in \mathbb{R}^{|N|} \ \middle| \ \sum_{i \in N} \pi_i = \overline{c}^{IV}(N) \text{ and } \sum_{i \in S} \pi_i \le \overline{c}^{IV}(S) \text{ for all } S \subset N, \ S \ne \emptyset \right\}. \tag{3.19}$$

$\underline{C}^{IV}(N, \underline{c}^{IV})$ and $\overline{C}^{IV}(N, \overline{c}^{IV})$ define the cores of the cooperative games $(N, \underline{c}^{IV})$ and $(N, \overline{c}^{IV})$, respectively; i.e., one core for the upper bounds and another one for the lower bounds. If we choose $(\underline{I}_1, \ldots, \underline{I}_{|N|}) \in \underline{C}^{IV}(N, c)$ and $(\overline{I}_1, \ldots, \overline{I}_{|N|}) \in \overline{C}^{IV}(N, c)$, then it follows immediately from the definitions of c^{IV}, I_i, and $C^{IV}(N, c^{IV})$ that $([\underline{I}_1; \overline{I}_1], \ldots, [\underline{I}_{|N|}; \overline{I}_{|N|}]) \in C^{IV}(N, c^{IV})$. Alparslan-Gök et al. (2008a) state that the time complexity of the algorithm for computing the interval core is the same as the time complexity of a related algorithm for computing the classical core (see Remark 3.1 in Alparslan-Gök et al. 2008a). Therefore, it can be concluded that an element in the interval core can easily be computed if one can compute an element in the core. To find core elements in the two games $(N, \underline{c}^{IV})$ and $(N, \overline{c}^{IV})$, the row generation procedure needs to be applied to both games successively. In Chap. 6, we will present an application for interval-valued games and show that above suggestion might bring up some interpretation problems.

3.7 Conclusions

In this chapter, a general approach to compute core cost allocations was introduced based on a constraint generation procedure.

Remarkably, this row generation procedure is not confined to a special problem. In the following chapters, we will present applications to several lot sizing problems (Chap. 5: economic lot sizing problem, Chap. 6: economic lot sizing problem with uncertain demand, Chap. 7: capacitated lot sizing problem, Chap. 8: multilevel lot sizing problem). Drechsel and Kimms (2010a) explain that the idea is basically valid for every game where the characteristic function is evaluated by solving an optimization problem – the subproblem may also be a non-linear mixed-integer model.

Drechsel and Kimms (2010a) discuss the influence of the subproblem as well: Depending on the model type of the subproblem, it will require different solution procedures. Nevertheless, this does not influence the proposed row generation algorithm because it does not rely on a specific algorithm for the subproblem. Contrary, the run-time performance is not equally good for all games – finding a core element is $\mathcal{NP}$-hard in general. Hence, solving the subproblem might be difficult in general. This is not due to the proposed approach, but due to the problem itself. Even in the definition of the core (2.7), the values $c(S)$ for the characteristic function are necessary parameters. Hence, there is no way to avoid computing them – but in contrast to the general definition, our procedure usually requires to compute just a few, but not all $c(S)$ values (see the computational studies in the following chapters: Sects. 5.3, 6.4, 7.4, and 8.3).

If the optimization problems become too complex and problem instances of practical size should be solved, the subproblem probably has to be solved by heuristics. For the following applications, the subproblems are hard to solve theoretically, but standard software solved them successfully in adequate time. If it is not possible to compute optimal values for $c(S)$, Drechsel and Kimms (2010a) suggest to use the best heuristical result. As the core definition does not require optimal solutions – it requires the "best" value a coalition S can reach. In such cases, we have to implement a heuristic to solve the subproblem during the course of our routine. This shows that even hard to solve optimization problems do not contradict the suggested approach.

Some difficulties when dealing with heuristics during the row generation have to be noticed: It may happen that a game is subadditive while solving the subproblem optimally but it is not when the $c(S)$ values are heuristic values (i.e., they are upper bounds if the underlying problem is a minimization problem). Drechsel and Kimms (2010a) suggest to use a pragmatic approach for cases where subadditivity is proved with regard to optimum $c(S)$ values: Whenever nonempty, disjoint player sets $S_1, \ldots, S_m$ appear while solving the subproblem heuristically which demonstrate that they will be better off acting alone; i.e., $c(S_1) + \cdots + c(S_m) < c(\bigcup_{k=1}^{m} S_k)$ where $c(\cdot)$ are heuristic values, replace $c(\bigcup_{k=1}^{m} S_k)$ in the right hand side of the specific stability constraint with $c(S_1) + \cdots + c(S_m)$. Then restart the procedure of computing a core element again with the modified master problem. We can simply explain this proceeding as that someone found a new (better) heuristic solution for the problem with the players $\bigcup_{k=1}^{m} S_k$.

Chapter 4
Cooperation in Supply Chains

Due to the growing number of supply chain management success stories, companies are more and more connected nowadays. Business units are forced to think in terms of complex supply networks rather than in terms of isolated decision making. Setting up a cooperation to improve the own performance is well established.

This chapter aims to identify decision problems in supply chains which can be influenced not only by a single decision maker but also via cooperations. Firstly, we will motivate why cooperations may be useful and recommendable in supply chains and discuss different forms of cooperation. Afterwards, we will review the literature regarding so-called "supply chain games" – cooperative decision problems in supply chains that could be solved with the help of cooperative game theory. These explanations shall be the basis for the developed cooperative games in supply chains following in the next chapters.

The research in the field of supply chain management already started in the 1980s; e.g., Simchi-Levi et al. (2004, p. 7) and de Kok and Graves (2003, p. 2) give an overview of the main trends that have induced supply chain management research. Literature is full of varying definitions of supply chain management and the aim of this chapter is not to discuss and evaluate those definitions. For that purpose, we refer to standard textbooks like, for instance, Simchi-Levi et al. (2004) and use their definition (see Simchi-Levi et al. 2004, p. 2): "Supply chain management is a set of approaches used to efficiently integrate suppliers, manufacturers, warehouses, and stores so that merchandise is produced and distributed at the right quantities, to the right locations, and at the right time in order to minimize costs within the system while satisfying service-level requirements." Although the term "supply chain" does not seem to be adequate because suppliers, manufacturers, and warehouses are mostly organized in networks and not only in linear chains, this term is commonly used in the literature rather than terms like "supply network", etc. Hence, we will use the term "supply chain" even though a network structure will not be denied. Most of the definitions that can be found in the literature have in common that supply chains are based on cooperation in order to generate benefits (see Thun 2005, p. 478). We will go into more detail regarding forms of cooperation in the next section.

Furthermore, the notion of supply chain management covers strategic, tactical, and operational management issues or decision-making processes arising from

J. Drechsel, *Cooperative Lot Sizing Games in Supply Chains*, Lecture Notes
in Economics and Mathematical Systems 644, DOI 10.1007/978-3-642-13725-9_4,
© Springer-Verlag Berlin Heidelberg 2010

managerial and economic considerations. Kogan and Tapiero (2007) see supply chain management as an important alternative to centralized and authoritarian-based approaches that are commonly used in management.

Apart from the manifold challenges arising from supply chain situations, the problems of supply chain management provide a large application area for operations research methods and techniques. In Sect. 4.2, we will show that cooperative game theory has already entered supply chain management.

4.1 Horizontal versus Vertical Cooperation

Cooperation can be defined as the process of coordinating goals and actions of agents; i.e., the participating partners should be coordinated (see e.g., Thun 2005). Buckley and Casson (1988) identify cooperation as a special type of coordination. For a comprehensive study regarding cooperation in the private sector, compare the work of Contractor and Lorange (1988a) and the therein contained contributions of Contractor and Lorange (1988b), Buckley and Casson (1988), and Harrigan (1988). For cooperation in public management, see e.g., Leach (2006). The notion *collaboration* is often used as a refinement for cooperation: Monczka et al. (2002, p. 135) define collaboration as "the process by which two or more parties adopt a high level of purposeful cooperation to maintain a trading relationship over time. The relationship is bilateral; both parties have the power to shape its nature and future direction over time. Mutual commitment to the future and a balanced power relationship are essential to the process." See also Völker and Neu (2008, pp. 46 and 76) for a more detailed discussion in which they also present special definitions for the term supply chain collaboration; e.g., "supply chain collaboration is often defined as two or more chain members working together, making joint decisions, and sharing benefits which result from greater profitability of satisfying end customer needs than acting alone" (see Simatupang and Sridharan 2002, p. 19, 2005, p. 45). Stadtler (2009) develops a comprehensive framework for collaborative planning. Apart from presenting the roots of collaborative planning, he builds up the framework on the following properties: structure of the supply chain, relationships among the partners, criteria of the underlying decision problem, and the characteristic of the planning scheme. In the sequel of this thesis, we will use the terms cooperation and collaboration equally.

From an organizational point of view, cooperations can be divided into *intraorganizational* and *interorganizational*; i.e., cooperating business units belonging all to the same company/organization versus cooperation of economically and legally independent companies.

The literature regarding such intercompany or intracompany cooperations distinguishes furthermore between horizontal cooperation (buyer-buyer, seller-seller) and vertical cooperation (buyer-seller). Examples for horizontal cooperation may be shared service centers, horizontal alliances, and horizontal cooperative procurement while vertical cooperation contains concepts like co-makership, vertical alliances, and public-private partnerships (see Schotanus 2007, p. 14). Schotanus (2007, p. 14) mentions the number of cooperating organizations as a big difference between horizontal and vertical cooperation because the number of partners in vertical technical alliances is rather low and the focus is more on technical capabilities. For economies of scale, the number of partners plays an essential role. Furthermore, Schotanus (2007) calls the cooperation theme a significant difference between horizontal and vertical cooperations: The first focuses on purchasing themes (e.g., quantity discounts), whereas the second one concentrates on improving processes and new technologies. Apart from those differences, many contributions on alliances can be taken into account when dealing with either horizontal or vertical cooperation.

Thun (2005, p. 478) states that the *vertical cooperation* builds the basis of supply chain management because companies from different supply chain levels or different industries cooperate. For instance, suppliers and manufacturers along a supply chain cooperate with the aim to work against the bullwhip effect. In general, every partner is responsible for specific production steps where overlapping is rather uncommon. Apart from that, vertical cooperation inside one company might exist if business units like procurement and research & development cooperate while developing new products. The most prominent reasons for a vertical cooperation are a decreasing vertical integration, securing of quality and quantity regarding the resources, decreasing raw material costs, and decreasing costs due to less organization in a cooperation (compare Arnold and Eßig 1997, p. 10). Brandenburger and Harborne (1996) offer an exact definition of the "added-value" which can be created by companies when organizing themselves in vertical cooperations: The created value equals the customer's willingness-to-pay minus the supplier's opportunity costs. Furthermore, they develop value-based strategies how to capture value.

Horizontal cooperations are coined by the joint work of companies that belong to the same supply chain stage and normally produce/trade the same products (they are competing and are part of the same industry). Contrary to vertical cooperation, where the companies complement one another, the firms in horizontal cooperations add their strength to act with more power compared to suppliers or customers (e.g., being a so-called "A-customer" instead of a "B-customer"). Accordingly, one special horizontal cooperation is the so-called purchasing alliance (cooperative purchasing) which can be found under several different names in the literature – Schotanus (2007, p. 12) list more than 150 of them. As it is common in the literature, we will use the terms purchasing and procurement equally.

The most prominent reasons for horizontal cooperation are to use synergistic effects (economies of scale, scope, and speed), reduce transaction costs due to a decreasing number of transactions, and, as already mentioned, strengthening the competitiveness. These advantages should not be compensated by drawbacks like increasing complexity of the purchasing process or the loss of flexibility and control. Schotanus (2007) compares advantages and disadvantages more or less to those of coordinated or centralized solution approaches (see Schotanus 2007, p. 13). The procurement's strategic and tactical impact can be explained with the help of Porter's basic model (see Porter 1985, pp. 37 and 41). Furthermore, caused by a decreased vertical integration during the last years, the weight of purchasing has increased dramatically. Thus, a more efficient procurement due to cooperation can enhance the company's success.

We refer to Schotanus (2007, pp. 15 and 26) for a literature survey regarding cooperative procurement. Apart from the work of Schotanus (2007), there are several other authors discussing problems of cooperative procurement: e.g., Johnson (1999), Essig (2000), Heijboer (2002), and Nagarajan et al. (2008). In Hendrick (1997) and later on in Schotanus (2007), extensive studies about the current utilization of consortium purchasing, consortium structure, legal aspects, success defining factors, purchased products, cost savings, etc., are provided. During his qualitative study, Schotanus (2007) found out that 87% of the purchasing groups uses the equal price gain allocation method. In Schotanus et al. (2008), the authors analyze unfairness resulting from using this allocation method and make suggestion to do better. Cruijssen et al. (2007) present results regarding a large survey on the potential benefit of horizontal cooperation under logistics service providers. The survey discloses that finding reliable partners and set up a fair allocation mechanism are the biggest obstacles.

For public organizations, horizontal cooperation in form of cooperative purchasing seems to be an interesting concept because typically, there is no competition between them. Hence, there is no incentive to outperform partners and no apprehension concerning confidential information. Furthermore, it is easy to define common interests and goals in such an alliance. This is an advantage because Schotanus (2007, p. 25) calls these factors two of the most important ones for an efficient cooperation. In addition, small and medium sized companies have big interests into horizontal cooperation to enhance their bargaining power compared to their suppliers (see Arnold 1998 for examples regarding cooperative procurement in Germany).

Strategic alliances are an important item in the context of supply chain cooperation as well. Simchi-Levi et al. (2004, p. 112) define strategic alliances as "multifaceted, goal-oriented, long-term partnerships between two companies in which both risks and rewards are shared". Compare also Todeva and Knoke (2005), Völker and Neu (2008, p. 78), and Shenkar and Reuer (2006) for more insights into strategic alliances in the context of collaboration and Pyke and Johnson (2004) for using strategic alliances as sourcing strategies. Lewis (1990) develops a general framework for analyzing strategic alliances. He focuses on the aspect of gaining competitive advantage. For strategic alliances in supply chains, compare Simchi-Levi et al. (2004, p. 111), where also concrete strategic alliances, like third-party logistics, retailer-supplier partnerships (e.g., vendor managed inventory), or distributor integration, are discussed. Whether or not a strategic alliance arises as vertical or horizontal cooperation is not clearly separated in the literature (see the above definition).

The supply chain situations which we will study in Chaps. 5–7 belong to the field of horizontal cooperation. The economic lot sizing game (introduced in Chap. 5) describes a situation where several companies or business units purchase raw materials or products together. The capacitated lot sizing game (see Chap. 7) deals with cooperative production, but again on the same supply chain level. The last model, discussed in Chap. 8, includes multilevel supply structures and thus contains aspects of horizontal as well as vertical cooperation.

4.2 Supply Chain Games in the Literature

The connections between single companies or business units in a supply network – independent of contracts or other binding agreements – directly lead to a multi-decision maker framework. Thus, game theory seems to provide an adequate modeling basis for problems in supply chain management. Cachon and Netessine (2004), Leng and Parlar (2005), Meca and Timmer (2008), and Nagarajan and Sošić (2008) give recent surveys about game theoretic approaches and applications in supply chain management. Sometimes this topic also appears under the notion supply chain collaboration. Kogan and Tapiero (2007) present deep insights into many supply chain games for discrete-time and continuous-time models.

There are several contributions dealing with cooperation in supply chains and focusing on different lot sizing problems like the economic order/production quantity, the newsvendor, or the Wagner–Whitin problem. Meca et al. (2004) develop the term inventory game for a cooperative procurement situation. Starting with the simplest inventory model – the *economic order quantity* (EOQ) where demand is continuous over time and at a constant rate – they design a scenario with several firms or shops (e.g., franchise operators) and a single product. Such an EOQ ordering situation can be modeled in closed form. Hence, this approach has the advantage that the problem can be treated analytically and they derive results as a closed form solution. In Meca et al. (2003), these results are consigned to the *economic production quantity* (EPQ) game with shortages. They prove that the EPQ game with shortages leads to exactly the same class of cost games as their earlier introduced EOQ game. Furthermore, they develop a proportional-like cost sharing rule (Share the Ordering Costs – SOC) for both of the games (see also Mosquera et al. 2008 for a note on this specific topic). Meca (2007) continues this work for generalized holding cost games where the holding costs h_S for every coalition S are assumed as $h_S := \min_{j \in S}\{h_j\}$ (i.e., products are stored in the warehouse of the player with lowest holding costs). The EPQ in combination with temporary discounts is studied by Meca et al. (2007). A setting with a multi-product EOQ is investigated in Dror and Hartman (2007). They reformulate the problem using a new matrix framework, which can be used to establish conditions when order consolidation has a nonempty core. Minner (2007) develops a joint replenishment model based on the EOQ, presents solution approaches for short-term and long-term cooperative sourcing, and analyzes a Nash bargaining solution.

Wagner–Whitin-related games consider deterministic and discrete dynamic demand over multiple periods (for the classical Wagner–Whitin problem see Wagner and Whitin 1958). Guardiola et al. (2009) call them production inventory games where unit costs for holding, backlogging, and production are taken into account – in contrast to fixed ordering costs. They prove that this game is totally balanced and that an element in the core can be computed in polynomial time. The research concerning those games is extended in Guardiola et al. (2008) regarding characterizations of the Owen point. Fixed ordering costs are added in the work of Guardiola et al. (2006) – they call it the setup inventory game. For such a setup inventory game, Chen and Zhang (2006) use linear programming duality to compute a core element in polynomial time. This approach heavily relies on a specific model formulation. Another contribution to the topic of setup inventory games is presented by van den Heuvel et al. (2007a). They use the Wagner–Whitin model as basis and call it economic lot sizing (ELS) game. A very recent contribution is the paper of Xu and Yang (2009), who develop a specific algorithm to compute cost shares but does not focus on the core properties.

For a cooperative ordering situation with *stochastic demand*, Hartman et al. (2000) prove that under specific assumptions (e.g., for any combination of normally distributed individual demands), the core is not empty. Such a cooperative game based on a newsvendor setting is called inventory centralization game in the literature. Slikker et al. (2001) and Özen et al. (2010) study newsvendor games and their behavior regarding convexity and emptiness of the core as well. Finally, Müller et al. (2002) prove that the core of a newsvendor game is always nonempty for all possible joint distributions of the random demands. Hartman and Dror (2003) develop a greedy optimization procedure to compute a solution that minimizes the costs of centralization. Apart from computing solutions where the players are better off, they provide an analysis on how to maximize their collective benefits. In the following work, Hartman and Dror (2005) present conditions regarding holding and penalty costs for assuring subadditivity and a nonempty core of the newsvendor game. The newsvendor problem can be examined under dynamic aspects; i.e., demand realizations over several periods. Such an environment is considered by Dror et al. (2008). Even though this game has, in general, an empty core, a repeated cost allocation scheme is provided. Another conceivable extension of the newsvendor game considers transshipments between the players. Slikker et al. (2005) examine this variant. Özen et al. (2007) include a warehouse in-between the producer and the retailer. Demands are realized when the goods reach the warehouse and order quantities can be reallocated before shipping from the warehouse to the retailer takes place. They consider two contracting schemes (buy-back and wholesale-price contract) and prove the non-emptiness of the core for such a game. A newsvendor situation with restriction on the minimum delivery quantity is studied by Özen et al. (2006) – a discussion regarding the consequences on the profit sharing and some new monotonicity properties are included. Some complexity results are derived by Chen and Zhang (2007). They show that determining whether an allocation is in the core of a newsvendor game is $\mathcal{NP}$-hard even in very simple settings. Chen (2009) introduce price-dependent demand and quantity discount aspects into the newsvendor environment. They prove the non-emptiness of the core and show how to compute an allocation in the core. Guardiola et al. (2007) study newsvendor games with quantity discounts as well. To increase expected joint profits, reallocation of the orders among the retailers could make sense. Such demand updates are discussed by Özen and Sošic (2006). Özen et al. (2008) develop a game with several newsvendors and warehouses and prove that the core is nonempty for such games. Montrucchio and Scarsini (2007) develop large newsvendor games, i.e., the number of participating newsvendors is infinite. Due to this fact, the newsvendor game becomes a nonatomic game. Discussions regarding balancedness and uniqueness of the core are included.

Gerchak and Gupta (1991) use stochastic demand for an inventory centralization setting as well, but under continuous review. They show that cooperation (centralization) is economical and investigate different methods of cost allocation regarding their applicability. Hartman and Dror (1996) continue the studies in this field and introduce three properties (stability, justifiability, and polynomial computability) that should hold for the cost allocation if the centralized system should last. For the joint replenishment problem, Zhang (2009) shows that this game has a nonempty core under power-of-two policies, provided that the joint setup cost function is submodular.

Apart from those games concentrating on different ordering situations, there are other applications in the field of supply chains like, e.g., *transportation* and *vehicle routing*. Özener and Ergun (2008) study a collaborative transportation procurement network where shippers cooperate for a higher utilization of truck capacity. Krajewska et al. (2008) develop a cooperative game based on horizontal cooperation among freight carriers – the game combines aspects of scheduling and routing. They choose the Shapley value to determine fair allocations and present results regarding real-life instances. A vehicle routing game is also examined by Göthe-Lundgren et al. (1996) who develop an algorithm to compute the nucleolus for such games. Engevall et al. (2004) extend this vehicle routing game to a heterogeneous vehicle fleet and apply it to a real-life case in a gas company.

Furthermore, there are several contributions regarding *location games*: An early example is the paper of Tamir (1992). He studies situations where several demand points (customers) should be served from centers and how to allocate the costs of establishing the centers among the customers. Tamir's work shows that there is a class of location games for which the core is not empty. A distribution center consolidation game is examined by Klijn and Slikker (2005). The basic idea includes that several distribution centers cooperate and reassign their shipments and stocks in the cooperation to reduce total costs. It can be shown that the core is not empty for special demand distributions.

Chapter 5
An Economic Lot Sizing Game

The following section will introduce a practical cooperative problem occurring in supply chain situations: Cooperation in placing orders. This is not only interesting for private companies but also for public bodies who already join cooperative procurement programs to form a purchasing alliance and to benefit from economies of scale (see Sect. 4.1).

Note, we will use the terms procurement problem, ordering problem, and lot sizing problem as synonyms in the following.

5.1 Cooperative Ordering Situations

5.1.1 The Underlying Problem

The starting point is the basic economic lot sizing problem introduced by Wagner and Whitin (1958) – also known as the Wagner–Whitin problem. A single decision maker (a producer or retailer) is faced with demand d_t for a single item over the time horizon T. To satisfy this demand without backlogging and without shortages in every period, he has to make order decisions. Demand can be ordered in the same period where it is needed or in periods before. If the order is placed before, then the ordered items must be stored until the demand occurs. We search for the ordering quantity q_t in a way that total costs are minimized. The total costs consist of holding costs c_t^h per stored item, fixed ordering costs c_t^s (incurs whenever an order is placed in period t) and unit ordering costs c_t^p. The decision maker faces the classical trade-off between fixed costs versus holding costs – a so-called lot sizing problem occurs. Depending on the order quantities and the demand, there are I_t units of the item on stock at the end of period t (with I_0 giving the starting inventory at the beginning of period 1, w.l.o.g. $I_0 = 0$). The described problem can be formalized as a mixed-integer programming problem:

$$\min \sum_{t=1}^{T} (c_t^p q_t + c_t^s x_t + c_t^h I_t) \tag{5.1}$$

J. Drechsel, *Cooperative Lot Sizing Games in Supply Chains*, Lecture Notes in Economics and Mathematical Systems 644, DOI 10.1007/978-3-642-13725-9_5, © Springer-Verlag Berlin Heidelberg 2010

s.t.

$$
\begin{align}
I_t &= I_{t-1} + q_t - d_t & t &= 1, \ldots, T & (5.2)\\
q_t &\le M\,x_t & t &= 1, \ldots, T & (5.3)\\
I_t,\, q_t &\ge 0 & t &= 1, \ldots, T & (5.4)\\
x_t &\in \{0, 1\} & t &= 1, \ldots, T. & (5.5)
\end{align}
$$

Decision variables in this model are the order quantity q_t, in which period an order is placed ($x_t = 1$ if an order in period t is placed, 0 otherwise), and the inventory level I_t. M is a sufficiently big number (e.g., $M = \sum_{t=1}^{T} d_t$). The objective function (5.1) minimizes the sum of fixed and quantity dependent ordering costs as well as holding costs. Constraints (5.2) are inventory balances for every period so that at the end of every period the inventory level includes the starting inventory of this period plus the quantity of ordered products in this period minus the period's demand. Restrictions (5.3) assure that whenever the ordered quantity is positive, the indicator variable x_t is set to one. Equations (5.4) and (5.5) specify the domain of the decision variables. The restriction to non-negative I_i-values assures that no backlogging or shortages occur.

The presented Wagner–Whitin problem (5.1)–(5.5) can be solved very efficiently. However, it is still an optimization problem where no closed formula can be provided to receive optimal results. Compare the contributions of Federgruen and Tzur (1991), Wagelmans et al. (1992), and Aggarwal and Park (1993) for efficient solution procedures concerning the Wagner–Whitin problem.

The described situation assumes a single decision maker (we call it player in the following). Now, imagine that multiple players think of a cooperation. Let the set of players be N and $|N| \ge 2$ in the case of cooperation. All players face the same ordering situation – they are ordering the same item from the same supplier and cooperation means that these players decide to place orders together as one. Formally, each player $i \in N$ has a demand d_{it} that should be met in period t. If all players of set N cooperate, they face a joint demand

$$
d_t(N) = \sum_{i \in N} d_{it} \qquad t = 1, \ldots, T.
$$

The incentive for cooperative ordering is to reduce costs; in particular, fixed costs can be lowered. Players ordering together are called a coalition. Basically, all subsets $S \subseteq N$ are feasible coalitions. We assume equal setup and holding costs for each player, hence, we can just sum up the individual demands for the coalitions. Following van den Heuvel et al. (2007a), the problem to be solved for this group of players can be couched as:

$$
c(N) = \min \sum_{t=1}^{T} (c_t^p q_t + c_t^s x_t + c_t^h I_t) \qquad (5.6)
$$

s.t.

$$I_t = I_{t-1} + q_t - d_t(N) \qquad\qquad t = 1, \ldots, T \qquad (5.7)$$

$$q_t \le M\, x_t \qquad\qquad\qquad\qquad t = 1, \ldots, T \qquad (5.8)$$

$$I_t,\, q_t \ge 0 \qquad\qquad\qquad\qquad t = 1, \ldots, T \qquad (5.9)$$

$$x_t \in \{0, 1\} \qquad\qquad\qquad\qquad t = 1, \ldots, T. \qquad (5.10)$$

$c(N)$ represents the total costs for ordering jointly in the grand coalition N. By construction, total costs for an arbitrary coalition S are positive ($c(S) \ge 0$ for all $S \subseteq N$) and total costs for an empty coalition equals 0 ($c(\emptyset) = 0$).

It is obvious that there is no need for a new solution method to solve the *cooperative economic lot sizing problem* (ELS game) as stated in (5.6)–(5.10). Although the players cooperate, they are still independent companies or business units. Hence, the question arises how to allocate the total costs to the single members of the grand coalition. Before we will tackle this question in Sect. 5.2, we will investigate the properties of the ELS game.

5.1.2 Properties of the ELS Game

The most important properties to characterize cooperative games were already introduced in Sect. 2.2.2. Since we consider costs (money) for the ELS game, this game is a TU game.

The ELS game is *monotone* if (2.1) (see p. 11) holds: Due to $d_t(S_1) \le d_t(S_2)$ for $S_1 \subseteq S_2 \subseteq N$, the characteristic function of the ELS game is monotone.

For *subadditivity*, we have to check whether (2.2) (see p. 12) holds: Combining the optimal solutions of (5.6)–(5.10) for S_1 and S_2 ($S_1, S_2 \subseteq N, S_1 \cap S_2 = \emptyset$) yields a feasible solution for $S_1 \cup S_2$. For this reason, the optimal solution for $S_1 \cup S_2$ is at all times smaller or equal to the sum of the costs for S_1 and S_2. Guardiola et al. (2006) derive this result as a consequence of the totally balanced character of this type of games. Therefore, the ELS game is subadditive and the players have an incentive to form the grand coalition N.

Furthermore, we can determine *concavity* for the ELS game by (2.3) (see p. 13): van den Heuvel et al. (2007a) show that this game is not concave in general (they provide a counterexample, see Sect. 5.2.2). Nevertheless, for two special cases they can prove concavity: The two-period case ($T = 2$) and the case where all players face equal demand ($d_{it} = d_{jt}$ for all $i, j \in N$ and $t = 1, \ldots, T$).

5.2 Computing Core Cost Allocations for the ELS Game

It is verified by van den Heuvel et al. (2007a) that the ELS game has a nonempty core. However, they do not provide a way to compute a core element for the general case. Due to the possible non-concavity of the ELS game, marginal vectors cannot be used to compute core allocations. Therefore, the ELS game is a good application for the before proposed row generation procedure (see Sect. 3.2). Before running the procedure, we need to specify Step 4 of the algorithm (see p. 44) – finding a coalition that violates relaxed stability constraints.

5.2.1 The Row Generation Procedure

Drechsel and Kimms (2010a) clarify the open question how to find a coalition $S' \notin S$ ($S' \neq \emptyset$) that violates a stability constraint that is not considered up to this point in the procedure ($\sum_{i \in S'} \pi_i > c(S')$). To describe such a coalition mathematically, we introduce binary decision variables z_i for all $i \in N$ that indicate whether or not a player i belongs to the coalition S' ($z_i = 1$ if $i \in S'$, 0 otherwise). We can build up a constraint satisfaction problem containing a restriction for finding coalitions S'. We call it the subproblem:

$SP(\pi)$:

$$I_t = I_{t-1} + q_t - \sum_{i \in N} d_{it} z_i \qquad\qquad t = 1, \ldots, T \qquad (5.11)$$

$$q_t \leq M\, x_t \qquad\qquad t = 1, \ldots, T \qquad (5.12)$$

$$\sum_{i \in N} \pi_i z_i \geq \delta + \sum_{t=1}^{T} (c_t^p q_t + c_t^s x_t + c_t^h I_t) \qquad\qquad (5.13)$$

$$I_t,\, q_t \geq 0 \qquad\qquad t = 1, \ldots, T \qquad (5.14)$$

$$x_t \in \{0, 1\} \qquad\qquad t = 1, \ldots, T \qquad (5.15)$$

$$z_i \in \{0, 1\} \qquad\qquad i \in N. \qquad (5.16)$$

The values for π_i are taken from the previous master problem. δ is an arbitrary small value. We assume w.l.o.g. that all parameter values are integer-valued. Hence, an optimum solution of $MP(S)$ exists with all π_i values being integral if $w = 0$. Hence, (5.13) is equivalent to $\sum_{i \in N} \pi_i z_i > \sum_{t=1}^{T} (c_t^p q_t + c_t^s x_t + c_t^h I_t)$ for $\delta = 1$. The remaining constraints are already known from the cooperative economic lot sizing problem. If no feasible solution exists for this problem, then there is no coalition S' violating any stability constraint. Otherwise, the values for z_i define a coalition for which a stability constraint should be considered in the master problem during the next iteration(s).

Note, $S' = N$ is not feasible due to the efficiency constraint in combination with (5.13). $S' = \{\}$ will not occur because then we would have $\sum_{i \in N} z_i = 0$ which contradicts (5.13).

Due to the lack of an objective function, this subproblem may find many different coalitions. Therefore, we add an objective function to make the subproblem an optimization problem. Several different objectives are possible. Let us start with searching for a coalition of smallest cardinality:

$SP'(\pi)$:

$$\min \sum_{i \in N} z_i$$

s.t.

$$(5.11)\text{–}(5.16).$$

The procedure can be specified as follows:

$Core'(MP, SP')$:

1. Define a small initial set $\mathcal{S}$; e.g., $\mathcal{S} = \{\{1\}, \{2\}, \ldots, \{|N|\}\}$. Compute the individual total costs $c(S)$ for those coalitions $S \in \mathcal{S}$ and the total costs $c(N)$ for the coalition N.
2. Solve the linear program $MP(\mathcal{S})$ (see (3.1)–(3.5)) optimally.
3. If $w > 0$, stop the algorithm because the instance has an empty core.
4. Otherwise, try to solve $SP'(\pi)$ optimally.
5. If $SP'(\pi)$ has no feasible solution, then stop the algorithm because the found allocation is in the core.
6. Otherwise, compute the total costs $c(S')$ for this coalition, update $\mathcal{S} = \mathcal{S} \cup \{S'\}$, and go to Step 2.

Due to the objective function, the algorithm starts finding small coalitions that violate the core. In every following iteration, the program needs to search only for bigger coalitions. We can refine this idea by eliminating coalitions which probably will not be in the optimum solution of a subproblem. Hence, we try to reduce the search space at least temporarily. The parameter γ denotes the size of a coalition found in the previous iteration ($\gamma = |S'|$). The starting value for γ is 2 because only the singleton coalitions are defined to be in the initial set $\mathcal{S}$.

$SP''(\pi)$:

$$\min \sum_{i \in N} z_i$$

s.t.

$$\sum_{i \in N} z_i \geq \gamma$$
$$(5.11)\text{–}(5.16).$$

We have to make some small modifications in the procedure:

$Core''(MP, SP'')$:

1. Initialize $\gamma = 2$ and $flag = 0$. Define a small initial set $\mathcal{S}$; e.g., $\mathcal{S} = \{\{1\}, \{2\}, \ldots, \{|N|\}\}$. Compute the individual total costs $c(S)$ for those coalitions $S \in \mathcal{S}$ and the total costs $c(N)$ for the coalition N.

2. Solve the linear program $MP(\mathcal{S})$ (see (3.1)–(3.5)) optimally.
3. If $w > 0$, stop the algorithm because the instance has an empty core.
4. Otherwise, try to solve $SP''(\pi)$ optimally.
5. If $SP''(\pi)$ has no feasible solution and *flag* $= 1$, then set *flag* $= 0$ and $\gamma = 2$ and return to Step 4. If $SP''(\pi)$ has no feasible solution and *flag* $= 0$, then stop the algorithm because the found allocation is in the core.
6. Otherwise, compute the total costs $c(S')$ for this coalition, update $\mathcal{S} = \mathcal{S} \cup \{S'\}$, update $\gamma = |S'|(= \sum_{i \in N} z_i)$, update *flag* $= 1$, and go to Step 2.

With a slight change, two more alternative subproblem variants can be formulated. This time, the algorithm searches for coalitions of largest size.

$\tilde{SP}'(\pi)$:

$$\max \sum_{i \in N} z_i$$

s.t.

$$(5.11)\text{–}(5.16).$$

$\tilde{SP}''(\pi)$:

$$\max \sum_{i \in N} z_i$$

s.t.

$$\sum_{i \in N} z_i \leq \gamma$$

$$(5.11)\text{–}(5.16).$$

Procedure $Core'(MP, \tilde{SP}')$ stays the same. For procedure $Core''(MP, \tilde{SP}'')$, we have to initialize $\gamma = |N| - 1$ instead of $\gamma = 2$ in Steps 1 and 5 because we are starting with big coalitions now.

Apart from guiding the search for a coalition S' with specific size, one could design a subproblem that looks for a coalition S' that violates the stability constraint $\sum_{i \in S'} \pi_i \leq c(S')$ most:

$\hat{SP}(\pi)$:

$$\max -o + \sum_{i \in N} \pi_i z_i$$

s.t.

$$o = \sum_{t=1}^{T} (c_t^p q_t + c_t^s x_t + c_t^h I_t)$$

$$(5.11),\ (5.12),\ \text{and}\ (5.14)\text{–}(5.16),$$

$$o \geq 0.$$

The new decision variable o denotes the characteristic function value $c(S')$. Hence, this value does not need to be calculated separately as for the previous subproblem formulations. If the optimum objective function value of this subproblem is positive, then a coalition S' has been found that should be considered in the master problem in the next iteration (due to $\sum_{i \in N} \pi_i z_i > o$). On the other hand, if the optimum objective function value is less than or equal to zero, the procedure can be terminated because the actual π_i values define a core element. In the contribution of van den Heuvel et al. (2007b), they prove that this subproblem is $\mathcal{NP}$-complete. The whole procedure is the following:

$\hat{Core}(MP, \hat{SP})$:

1. Define a small initial set $\mathcal{S}$; e.g., $\mathcal{S} = \{\{1\}, \{2\}, \ldots, \{|N|\}\}$. Compute the individual total costs $c(S)$ for those coalitions $S \in \mathcal{S}$ and the total costs $c(N)$ for the coalition N.
2. Solve the linear program $MP(\mathcal{S})$ (see (3.1)–(3.5)) optimally.
3. If $w > 0$, stop the algorithm because the instance has an empty core.
4. Otherwise, solve $\hat{SP}(\pi)$ optimally.
5. If $\hat{SP}(\pi)$ has a non-positive optimum objective function value, then stop the algorithm because the found allocation is in the core.
6. Otherwise, set $c(S') = o$, update $\mathcal{S} = \mathcal{S} \cup \{S'\}$ and go to Step 2.

When using the algorithm to compute an element in the ϵ-core or in the least core, the objective function of the subproblem needs to be replaced by

$$\max \sum_{i \in S'} \pi_i - c(S') - \epsilon.$$

For the least core, the value ϵ used in the subproblem equals the objective function value of the most recent master problem.

5.2.2 A Numerical Example

Table 5.1 displays the parameters for a small ELS game with three players and six periods to illustrate the working principle of the proposed procedure. This non-concave instance is taken from van den Heuvel et al. (2007a), Example 5.[1] We assume $I_0 = 0$. Table 5.2 presents the characteristic function values for all possible subcoalitions $S \subseteq N$. It can be seen easily that all players benefit from cooperating (e.g., $644 + 511 > 1{,}029$) and that this instance is not concave $(c(\{1, 2\}) - c(\{1\}) = 385 < 389 = c(\{1, 2, 3\}) - c(\{1, 3\}))$. An equal cost allocation among the three players would lead to $\pi_1 = \pi_2 = \pi_3 = 464.3$. Such an allocation would discriminate against players 2 and 3 because they would be better off cooperating without player 1 to receive $c(\{2, 3\}) = 869$.

---

[1] Note that this example contains misprinted parameter values in the original paper. Van den Heuvel provided us with the corrected values upon request.

Table 5.1 ELS game: parameters for the numerical example

t	1	2	3	4	5	6
d_{1t}	15	5	14	9	20	11
d_{2t}	1	1	11	17	3	14
d_{3t}	20	8	1	11	1	19
s_t	0	100	132	71	77	111
h_t	5	3	5	4	2	1
p_t	1	11	4	7	8	8

Table 5.2 ELS game: characteristic function values for the numerical example

S	$\emptyset$	$\{1\}$	$\{2\}$	$\{3\}$	$\{1, 2\}$	$\{1, 3\}$	$\{2, 3\}$	$\{1, 2, 3\}$
$c(S)$	0	644	511	483	1,029	1,004	869	1,393

The core of this game is represented by the following equations:

$$\pi_1 + \pi_2 + \pi_3 = 1,393$$
$$\pi_1 \leq 644$$
$$\pi_2 \leq 511$$
$$\pi_3 \leq 483$$
$$\pi_1 + \pi_2 \leq 1,029$$
$$\pi_1 + \pi_3 \leq 1,004$$
$$\pi_2 + \pi_3 \leq 869.$$

Figure 5.1 displays the graphical illustration of the core for this instance – the core for a three player game is a two-dimensional polyhedron.

When applying the proposed row generation algorithm, we want to achieve computing a core cost allocation without using all stability constraints. Recall that we start with the efficiency constraint and the stability constraints concerning the single coalitions.

Iteration 1: We solve the Wagner–Whitin problems optimally for the single player coalitions (values see Table 5.2) and solve the master problem $MP(\{1\}, \{2\}, \{3\})$ optimally:

$$\min w$$

s.t.

$$\pi_1 + \pi_2 + \pi_3 = 1,393$$
$$\pi_1 - w \leq 644$$
$$\pi_2 - w \leq 511$$
$$\pi_3 - w \leq 483$$
$$w \geq 0.$$

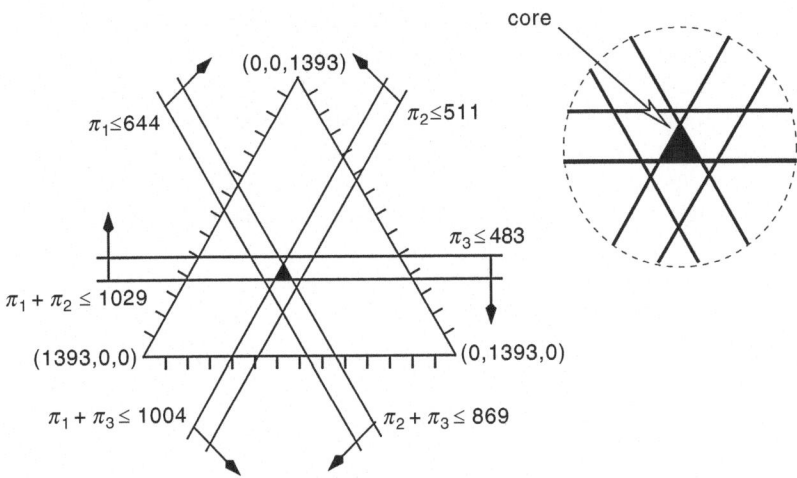

Fig. 5.1 ELS game: the core $\{(\pi_1, \pi_2, \pi_3)\}$ for the numerical example with $MP(\mathcal{S})$ (Drechsel and Kimms 2010a)

The optimum solution is $w = 0$ and $\pi = (644, 511, 238)$. Hence, the core is not empty and we continue the procedure to find a coalition violating stability constraints which were not taken into account. Solving the subproblem optimally results in $z_1 = z_2 = 1$ and $z_3 = 0$ which is obviously true because $\pi_1 + \pi_2 = 1,155 > 1,029 = c(\{1, 2\})$. That means, players 1 and 2 would have an incentive to form a smaller coalition than the grand coalition.

Iteration 2: We add the found coalition to $\mathcal{S}$ and solve $MP(\{1\}, \{2\}, \{3\}, \{1, 2\})$ optimally which means to solve the master problem from Iteration 1 with the additional stability constraint

$$\pi_1 + \pi_2 - w \le 1,029.$$

The optimum solution is again $w = 0$, but with $\pi = (518, 511, 364)$. We call the subproblem that gives the result $z_1 = 0$ and $z_2 = z_3 = 1$ because $\pi_2 + \pi_3 = 875 > 869 = c(\{2, 3\})$.

Iteration 3: Adding a further constraint to the master problem, we now solve $MP(\{1\}, \{2\}, \{3\}, \{1, 2\}, \{2, 3\})$ optimally which means to solve the master problem from Iteration 2 with the additional constraint

$$\pi_2 + \pi_3 - w \le 869.$$

The optimum solution is $w = 0$ and $\pi = (524, 505, 364)$. Calling the subproblem reveals that this solution is in the core because no solution with a positive objective function value can be found. Hence, the algorithm terminates. We can easily check whether this is true: We manually test the remaining stability constraint $\pi_1 + \pi_3 = 524 + 364 = 888 \leq 1{,}004 = c(\{1,3\})$ which holds.

Note that the found core allocation $\pi = (524, 505, 364)$ is an extreme point of the core. Depending on the solution procedure for the master problem (a linear program) we could get an extreme point, a vertex point, or an inner point of the core.

As stated in Sect. 2.3.4, the core of every cooperative TU game is a closed and bounded convex set. We can specify the core for our numerical example:

$$
\begin{aligned}
C(N,c) = \{ & \lambda_1(524, 505, 364) \\
 & +\lambda_2(524, 389, 480) \\
 & +\lambda_3(640, 389, 364) \mid \lambda_1 + \lambda_2 + \lambda_3 = 1 \text{ and } \lambda_1, \lambda_2, \lambda_3 \geq 0 \}.
\end{aligned}
$$

For comparison, the least core for this instance is the ϵ-core with $\epsilon = -38.67$ and

$$
C^L(c) = \{(562.67, 427.67, 402.67)\}.
$$

In Sect. 3.4, we described two alternative formulations for the master problem to control the choice of core elements regarding fairness. Replacing $MP(S)$ by $MP^I(S)$, we try to compute a core allocation for our numerical example that minimizes the differences between the three cost shares. We apply again the row generation procedure and the result is the cost allocation $\pi = (524, 434.5, 434.5)$. The solution is illustrated in Fig. 5.2.

If we replace $MP(S)$ by $MP^{II}(S)$ and apply the procedure to the example, the cost allocation is $\pi = (547.68, 434.57, 410.76)$ – a cost allocation that minimizes the relative differences between the cost shares compared to individual costs. The solution is illustrated in Fig. 5.3. In Sect. 3.4 (p. 48), we described the core

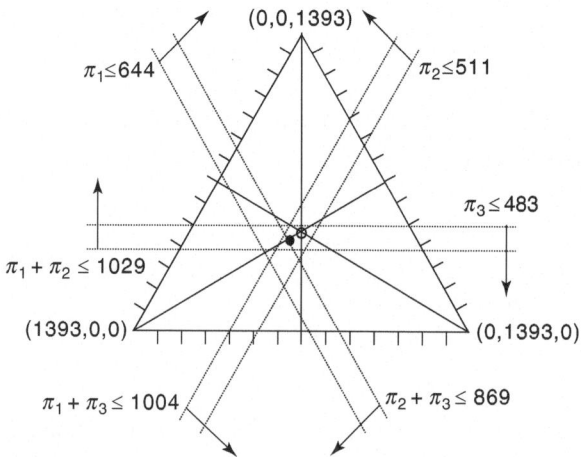

Fig. 5.2 ELS game: the core $\{(\pi_1, \pi_2, \pi_3)\}$ for the numerical example with $MP^I(S)$ (Drechsel and Kimms 2010a)

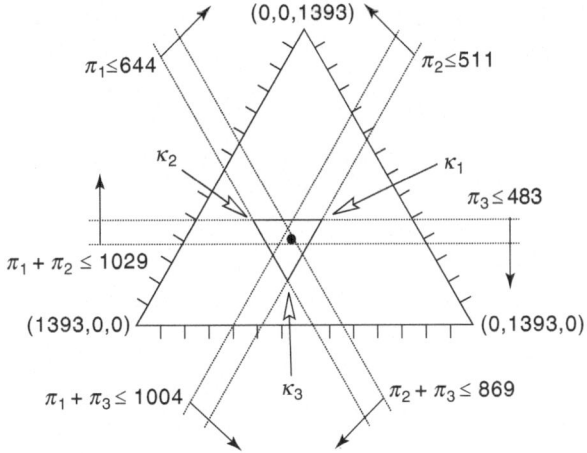

Fig. 5.3 ELS game: the core $\{(\pi_1, \pi_2, \pi_3)\}$ for the numerical example with $MP^{II}(\mathcal{S})$ (Drechsel and Kimms 2010a)

for $MP^{II}(\mathcal{S})$ as the convex combination of the $|N|$ vertices of the polyhedron $\{\pi \geq 0 | \sum_{i \in N} \pi_i = c(N)$ and $\pi_i \leq c(\{i\})$ for all $i \in N\}$. For our example, these vertices are $\kappa_1 = (399, 511, 483)$, $\kappa_2 = (644, 266, 483)$, and $\kappa_3 = (644, 511, 238)$ with the weights $\omega_1 = 0.39$, $\omega_2 = 0.31$, and $\omega_3 = 0.29$.

5.3 Computational Study for the ELS Game

The introduced algorithm was implemented using the commercial software AMPL/ CPLEX version 10.0.0. We conducted the tests on Intel Pentium hardware with 2.8 GHz and 504 MB RAM. Compare also Drechsel and Kimms (2010a) for the results regarding this computational study.

We determined the number of players $|N|$ as the key parameter for cooperative planning problems because the definition of the core and with it the complete master problem grows in the order of $2^{|N|}$. The run-time performance to find core cost allocations is very promising because even the largest instances in the test-beds were solved within reasonable time. For all instances with $|N| \leq 20$, we observed a run-time of only a few minutes. Accordingly, the run-time per iteration was fractions of a second for all instances with $|N| \leq 60$ and $T = 6$. Due to this fact, we will not report detailed results regarding run-time. In the following, we provide detailed information on the number of required iterations. This measure is independent of the used hardware and not influenced by the efficiency of the software implementation.

Apart from testing different instance characteristics (varying parameters), we studied the impact of the problem-specific part of the row generation procedure (i.e., the subproblem) on the performance. For this reason, we tested different versions of the subproblem described in Sect. 5.2. Recall, $\hat{SP}(\pi)$ searches for coalitions violating the stability constraints most, $SP'(\pi)$ and $SP''(\pi)$ find coalitions of smallest size and $\tilde{SP}'(\pi)$ and $\tilde{SP}''(\pi)$ find coalitions of largest size that violate the stability constraints. $SP''(\pi)$ ($\tilde{SP}''(\pi)$) additionally looks for a coalition size greater (smaller) than the size found in the previous iteration where the size counter is reset if no coalition violating the stability constraints can be found.

Test-Bed 1: For our basic instances, we generated random instances with $T = 6$ and parameter values being random integers which were drawn from the following intervals with uniform distribution: $d_{it} \in [0; 20]$, $c_t^s \in [0; 200]$, $c_t^h \in [0; 10]$, and $c_t^p \in [0; 15]$. The number of players was varied systematically:

$$|N| \in \{5, 10, 15, 20, 25, 30\}.$$

For each value $|N|$, 15 instances were generated. We terminated the procedure after 20,000 iterations in case that a core element has not been detected up to this point to keep the computational effort within reasonable limits. Table 5.3 displays the absolute number of iterations for the basic instances as an average over 15 instances (#*iter*) and the minimum and maximum number of iterations. The results reveal the efficiency of the approach. While the original core definition (2.7) requires $2^{|N|} - 2$ stability constraints, the row generation procedure used just $|N|$ constraints (from Step 1) plus (#*iter* − 1) generated restrictions on average. The percentage values in Table 5.3 are defined to be

$$\frac{(|N| + \#iter - 1) \times 100}{2^{|N|} - 2} \tag{5.17}$$

which is the portion of constraints that were actually used to determine the core allocation. We examined all variants of the master problem formulation ($MP(S)$, $MP^I(S)$, $MP^{II}(S)$) in combination with all variants of the subproblem formulation ($SP'(\pi)$, $SP''(\pi)$, $\tilde{SP}'(\pi)$, $\tilde{SP}''(\pi)$, $\hat{SP}(\pi)$).

For instance, Table 5.3 provides for the combination of $MP(S)$ with $SP'(\pi)$ for $|N| = 25$ the following results in our test with 15 instances: Six instances reached the iteration limit of 20,000 and were terminated before finding a core element. The average number of iterations over all 15 instances is larger than 9,063.8 (those six instances which were aborted contribute a value of 20,000 each to this average number). The above defined efficiency measure (5.17) reveals that this corresponds to an average of just 0.03% of generated constraints. At least one of the 15 instances already terminated after 24 iterations while the maximum number of iterations required is larger than 20,000 (in six cases).

Evaluating the results, we observe that the procedure is indeed efficient because only a small fraction of constraints is actually generated. A percentage value of 0.00% in Table 5.3 states that the percentage value being measured is smaller than 0.005%. As expected, the number of iterations grows faster than linear with a growing number of players, although the efficiency increases with an increasing number of players (the percentage values decrease). For instance, for $|N| = 30$,

Table 5.3 ELS game: average results test-bed 1 – average number of iterations / average percentage of required constraints (minimum/maximum number of iterations)

$\|N\|$	$T = 6$	$SP'(\pi)$	$SP''(\pi)$	$\tilde{SP}'(\pi)$	$\tilde{SP}''(\pi)$	$\hat{SP}(\pi)$
5	$MP(\mathcal{S})$	7.9/39.78	6.8/36.00%	5.6/32.00%	5.7/32.44%	4.0/26.67%
		(4/12)	(4/8)	(4/8)	(4/8)	(4/4)
	$MP^I(\mathcal{S})$	4.2/27.33%	4.1/26.89%	3.2/24.00%	3.2/24.00%	2.6/22.00%
		(2/10)	(2/8)	(2/8)	(2/8)	(2/4)
	$MP^{II}(\mathcal{S})$	1.1/17.11%	1.1/17.11%	1.1/17.11%	1.1/17.11%	1.1/17.11%
		(1/2)	(1/2)	(1/2)	(1/2)	(1/2)
10	$MP(\mathcal{S})$	139.7/14.55%	117.9/12.41%	17.5/2.60%	19.1/2.75%	10.5/1.91%
		(25/371)	(29/222)	(9/47)	(9/50)	(9/17)
	$MP^I(\mathcal{S})$	29.9/3.80%	23.7/3.20%	7.4/1.60%	7.5/1.61%	7.3/1.59%
		(11/61)	(10/38)	(3/20)	(3/18)	(3/10)
	$MP^{II}(\mathcal{S})$	3.5/1.23%	3.5/1.22%	1.9/1.07%	1.9/1.07%	1.9/1.06%
		(1/9)	(1/9)	(1/4)	(1/4)	(1/3)
15	$MP(\mathcal{S})$	1,159.1/3.58%	545.1/1.71%	24.3/0.12%	23.0/0.11%	15.3/0.09%
		(27/3,975)	(33/1,436)	(14/88)	(14/92)	(14/28)
	$MP^I(\mathcal{S})$	246.2/0.79%	155.4/0.52%	16.9/0.09%	15.1/0.09%	20.8/0.11%
		(56/407)	(48/268)	(7/92)	(7/67)	(9/33)
	$MP^{II}(\mathcal{S})$	44.3/0.18%	33.3/0.14%	5.7/0.06%	5.7/0.06%	7.7/0.07%
		(10/112)	(9/74)	(3/8)	(3/8)	(3/13)
20	$MP(\mathcal{S})$	^{2}>4,461.1/>0.43%	1,394.7/0.13%	315.9/0.03%	179.6/0.02%	61.8/0.01%
		(19/>20,000)	(20/7,782)	(19/2,416)	(19/1,220)	(19/519)
	$MP^I(\mathcal{S})$	982.9/0.10%	518.3/0.05%	65.5/0.01%	52.1/0.01%	56.7/0.01%
		(374/1,828)	(218/907)	(11/425)	(11/374)	(17/228)
	$MP^{II}(\mathcal{S})$	203.7/0.02%	139.5/0.02%	8.7/0.00%	8.7/0.00%	15.2/0.00%
		(9/628)	(8/424)	(4/12)	(4/12)	(5/40)
25	$MP(\mathcal{S})$	^{6}>9,063.8/0.03%	^{2}>7,455.1/0.02%	355.9/0.00%	298.7/0.00%	29.5/0.00%
		(24/>20,000)	(24/>20,000)	(24/1,489)	(24/1,317)	(24/75)
	$MP^I(\mathcal{S})$	2,647.0/0.01%	1,263.7/0.00%	575.6/0.00%	221.2/0.00%	131.4/0.00%
		(1,638/6,295)	(840/1,734)	(17/4,278)	(17/1,291)	(36/484)
	$MP^{II}(\mathcal{S})$	597.7/0.00%	342.5/0.00%	77.6/0.00%	41.1/0.00%	30.0/0.00%
		(187/1,231)	(122/750)	(8/630)	(8/281)	(12/63)
30	$MP(\mathcal{S})$	^{5}>7,886.3/0.00%	^{2}>6,960.6/0.00%	1,022.8/0.00%	463.1/0.00%	50.0/0.00%
		(35/>20,000)	(31/>20,000)	(29/11,954)	(29/4,343)	(29/169)
	$MP^I(\mathcal{S})$	6,452.4/0.00%	2,477.1/0.00%	420.3/0.00%	210.9/0.00%	192.8/0.00%
		(3,573/14,247)	(1,661/3,536)	(20/2,818)	(20/1,485)	(56/455)
	$MP^{II}(\mathcal{S})$	1,696.9/0.00%	975.5/0.00%	37.9/0.00%	31.5/0.00%	65.7/0.00%
		(308/2,819)	(226/1,443)	(11/302)	(11/188)	(42/84)

$2^{30}-2 > 10^9$ stability constraints exist. However, only about 66 stability constraints were generated on average for the combination $MP^{II}(\mathcal{S})/\hat{SP}(\pi)$.

A comparison of the three master problem formulations shows that the master problem $MP^{II}(\mathcal{S})$ has a better performance than the master problem $MP^I(\mathcal{S})$ independent from the chosen subproblem variant. The latter frequently performs better than the master problem formulation $MP(\mathcal{S})$.

If we compare the subproblem variants, we learn that searching for small coalitions ($SP'(\pi)$ and $SP''(\pi)$) performs much worse than its alternatives. This can be easily explained because adding constraints which affect only a few players tends to enforce smaller solution changes from iteration to iteration compared to adding constraints which affect many players. For a large number of players, the subproblem $\tilde{SP}''(\pi)$ outperforms $\tilde{SP}'(\pi)$ significantly. No clear dominance can be detected while comparing $\tilde{SP}''(\pi)$ and $\hat{SP}(\pi)$ in terms of average values when master problem $MP^{II}(S)$ is used. For the worst case, i.e., regarding the maximum number of required iterations, $\hat{SP}(\pi)$ delivers better values for most of the instances.

Test-Bed 2: Our basic instances in the first test-bed have fixed costs randomly drawn from the interval $c_t^s \in [0; 200]$. This does not depend on the number of players under consideration. Hence, for a growing number of players, the Wagner–Whitin problems for large coalitions tend to have an optimum solution which equals a lot-for-lot ordering policy; i.e., the order quantity in a period equals demand in that period because fixed costs are relatively low. Therefore, we set the fixed cost parameter depending on the number of players in the grand coalition: $c_t^s \in [100 \cdot (|N|/5 - 1); 100 \cdot (|N|/5 + 1)]$. The interval assures a width of 200, again. Thus, the Wagner–Whitin problems remain to be non-trivial even for a large number of players. All other parameter values were chosen as in Test-Bed 1. The number of instances per parameter constellation is 15, again.

Basically, the same observations as for the first test-bed can be made. Table 5.4 displays the results in the same manner like for Test-Bed 1. Again, combinations $MP^{II}(S)/\tilde{SP}''(\pi)$ and $MP^{II}(S)/\hat{SP}(\pi)$ have the best performance. It is interesting to note that the procedure terminates much faster compared with the results for Test-Bed 1 regarding the same combinations. For $|N| = 30$ and the combination $MP^{II}(S)/\hat{SP}(\pi)$, for instance, less than five constraints out of the $2^{30} - 2$ (10^9) are used on average. The reason seems to be that different coalitions have much more different costs now. Hence, the constraints which are actually needed to define the core can be detected much more easily as a look at Fig. 5.1 illustrates.

Test-Bed 3: Furthermore, we tested very large ELS instances to show that the proposed row generation procedure is also applicable if the underlying problem is very large. Hence, we tested instances of different size measured in terms of T. We used

$$T \in \{6, 12, 18, 24, 30, 36, 42, 48, 54, 100\}.$$

We chose the parameters like in Test-Bed 1 but investigated only instances with $|N| = 25$ players. Again, 15 random instances were used for each parameter setting. We only tested the two most promising master and subproblem combinations: $MP^I(S)/\hat{SP}(\pi)$ and $MP^{II}(S)/\hat{SP}(\pi)$. Table 5.5 presents the results. It is remarkable to observe that the row generation procedure tends to terminate earlier (after fewer iterations) the larger the optimization problem is. This is not strictly true, as Table 5.5 displays, but the trend is obvious. A reason for this behavior is not clear yet.

Table 5.4 ELS game: average results test-bed 2 – average number of iterations / average percentage of required constraints (minimum/maximum number of iterations)

$|N|$	$T = 6$	$SP'(\pi)$	$SP''(\pi)$	$\tilde{SP}'(\pi)$	$\tilde{SP}''(\pi)$	$\hat{SP}(\pi)$
5	$MP(\mathcal{S})$	7.9/39.78	6.8/36.00%	5.6/32.00%	5.7/32.44%	4.0/26.67%
		(4/12)	(4/8)	(4/8)	(4/8)	(4/4)
	$MP^I(\mathcal{S})$	4.2/27.33%	4.1/26.89%	3.2/24.00%	3.2/24.00%	2.6/22.00%
		(2/10)	(2/8)	(2/8)	(2/8)	(2/4)
	$MP^{II}(\mathcal{S})$	1.1/17.11%	1.1/17.11%	1.1/17.11%	1.1/17.11%	1.1/17.11%
		(1/2)	(1/2)	(1/2)	(1/2)	(1/2)
10	$MP(\mathcal{S})$	192.8/19.75%	123.2/12.94%	40.3/4.82%	29.3/3.75%	11.1/1.96%
		(25/511)	(21/288)	(9/128)	(9/76)	(9/31)
	$MP^I(\mathcal{S})$	8.8/1.74%	9.1/1.77%	5.5/1.42%	5.3/1.40%	3.5/1.23%
		(1/20)	(1/24)	(1/22)	(1/19)	(1/7)
	$MP^{II}(\mathcal{S})$	1.3/1.00%	1.3/1.00%	1.2/1.00%	1.2/1.00%	1.2/1.00%
		(1/4)	(1/4)	(1/3)	(1/3)	(1/3)
15	$MP(\mathcal{S})$	2,124.0/6.53%	877.4/2.72%	331.5/1.05%	142.7/0.48%	22.5/0.11%
		(66/9,289)	(37/2,118)	(16/1,507)	(16/423)	(14/89)
	$MP^I(\mathcal{S})$	47.6/0.19%	36.0/0.15%	11.5/0.08%	11.2/0.08%	7.6/0.07%
		(7/115)	(7/77)	(3/49)	(3/44)	(3/12)
	$MP^{II}(\mathcal{S})$	3.8/0.05%	3.9/0.05%	2.1/0.05%	2.1/0.05%	1.9/0.05%
		(1/16)	(1/14)	(1/4)	(1/4)	(1/3)
20	$MP(\mathcal{S})$	[6]>10,630.9/1.02%	[1]>6,124.2/0.59%	1,637.7/0.16%	651.5/0.06%	106.6/0.01%
		(27/>20,000)	(45/>20,000)	(19/10,996)	(19/2,502)	(19/728)
	$MP^I(\mathcal{S})$	166.7/0.02%	101.9/0.01%	79.1/0.01%	32.1/0.00%	15.0/0.00%
		(6/683)	(6/333)	(2/1,020)	(2/339)	(2/86)
	$MP^{II}(\mathcal{S})$	14.1/0.00%	13.1/0.00%	3.4/0.00%	3.5/0.00%	3.3/0.00%
		(1/70)	(1/72)	(1/6)	(1/6)	(1/7)
25	$MP(\mathcal{S})$	[12]>16,097.6/0.05%	[9]>14,814.2/0.04%	[2]>4,090.2/0.01%	2,142.4/0.01%	153.3/0.00%
		(24/>20,000)	(74/>20,000)	(24/>20,000)	(24/8,636)	(24/752)
	$MP^I(\mathcal{S})$	183.0/0.00%	128.8/0.00%	19.8/0.00%	17.7/0.00%	11.1/0.00%
		(28/605)	(26/379)	(4/190)	(4/159)	(4/26)
	$MP^{II}(\mathcal{S})$	20.6/0.00%	16.1/0.00%	3.2/0.00%	3.2/0.00%	3.6/0.00%
		(1/103)	(1/93)	(1/7)	(1/7)	(1/8)
25	$MP(\mathcal{S})$	[14]>18,676.3/0.00%	[8]>15,421.9/0.00%	4,552.7/0.00%	3,024.5/0.00%	120.3/0.00%
		(144/>20,000)	(148/>20,000)	(49/17,076)	(42/16,420)	(29/564)
	$MP^I(\mathcal{S})$	384.5/0.00%	244.5/0.00%	10.3/0.00%	10.4/0.00%	15.8/0.00%
		(2/1,440)	(2/676)	(2/37)	(2/40)	(2/38)
	$MP^{II}(\mathcal{S})$	49.7/0.00%	34.5/0.00%	4.0/0.00%	4.0/0.00%	4.3/0.00%
		(1/290)	(1/164)	(1/9)	(1/9)	(1/10)

Test-Bed 4: In the last test-bed, we investigated the impact of the number of players on the performance of the algorithm. We used

$$|N| \in \{30, 50, 100, 150\}$$

Table 5.5 ELS game: average results test-bed 3 – average number of iterations / average percentage of required constraints (minimum/maximum number of iterations)

$\lvert N \rvert = 25$	$MP^{I}(\mathcal{S})/\hat{SP}(\pi)$	$MP^{II}(\mathcal{S})/\hat{SP}(\pi)$
$T = 6$	131.4/0.00%	30.0/0.00%
	(36/484)	(12/63)
$T = 12$	145.7/0.00%	18.3/0.00%
	(39/1,027)	(3/34)
$T = 18$	89.1/0.00%	14.5/0.00%
	(43/356)	(5/46)
$T = 24$	124.5/0.00%	10.2/0.00%
	(21/612)	(2/30)
$T = 30$	68.3/0.00%	5.9/0.00%
	(19/140)	(3/16)
$T = 36$	70.4/0.00%	7.3/0.00%
	(28/201)	(1/22)
$T = 42$	46.6/0.00%	5.0/0.00%
	(12/116)	(1/21)
$T = 48$	40.6/0.00%	4.8/0.00%
	(19/57)	(2/14)
$T = 54$	41.1/0.00%	3.1/0.00%
	(17/85)	(1/8)
$T = 100$	17.7/0.00%	1.5/0.00%
	(6/50)	(1/3)

Table 5.6 ELS game: average results test-bed 4 – average number of iterations/average percentage of required constraints (minimum/maximum number of iterations)

$T = 6$	$MP^{I}(\mathcal{S})/\hat{SP}(\pi)$	$MP^{II}(\mathcal{S})/\hat{SP}(\pi)$
$\lvert N \rvert = 30$	192.8/0.00%	65.7/0.00%
	(56/455)	(42/84)
$\lvert N \rvert = 50$	501.3/0.00%	270.2/0.00%
	(140/2,555)	(109/1,074)
$\lvert N \rvert = 100$	2,763.8/0.00%	1,071.1/0.00%
	(772/7,083)	(689/1,351)
$\lvert N \rvert = 150$	[1]>3,751.0/0.00%	[1]>3,122.7/0.00%
	(1,735/>10,000)	(1,280/>10,000)

as well as the parameter constellations of our first test-bed. Likewise, 15 random instances per parameter setting were solved. We stopped the procedure if 10,000 iterations were run and no core allocation was found. As in Test-Bed 3, we only used the combinations $MP^{I}(\mathcal{S})/\hat{SP}(\pi)$ and $MP^{II}(\mathcal{S})/\hat{SP}(\pi)$. The results are presented in Table 5.6. As expected, the number of iterations grows if the number of players increases. However, we are still able to solve very large instances with adequate computational effort. Recall that for $\lvert N \rvert = 150$ players, $2^{150} - 2 > 10^{45}$ stability constraints exist while the algorithm needs less than 4,000 of them on average.

5.4 Extensions for the ELS Game

It should be emphasized that the described cost allocation approach is sufficiently general to apply it to other problems than the classical cooperative Wagner–Whitin problem, too. Let us exemplify some possible variants. To the best of our knowledge, these variants have not been studied as cooperative games in the literature. Hence, it is not clear up to now whether their core may be empty.

For the sake of simplicity, we assumed so far that all players face the same cost coefficients. It might happen that order unit costs or holding costs are player-dependent because each player uses own/different transportation possibilities and warehouses. To take this into consideration, we have to take parameters c_{it}^p and c_{it}^h into account. The following changes in the subproblem have to be made (the other parts of the procedure stay untouched):

$$\max -o + \sum_{i \in N} \pi_i z_i$$

s.t.

$$I_{it} = I_{i,t-1} + q_{it} - d_{it} z_i \qquad\qquad i \in N,\, t = 1, \ldots, T$$

$$\sum_{i \in N} q_{it} \le M\, x_t \qquad\qquad t = 1, \ldots, T$$

$$o = \sum_{t=1}^{T} \left(c_t^s x_t + \sum_{i \in N} (c_{it}^p q_{it} + c_{it}^h I_{it}) \right)$$

$$I_{it},\, q_{it} \ge 0 \qquad\qquad i \in N,\, t = 1, \ldots, T$$

$$x_t \in \{0, 1\} \qquad\qquad t = 1, \ldots, T$$

$$z_i \in \{0, 1\} \qquad\qquad i \in N$$

$$o \ge 0.$$

Apart from including player dependent cost coefficients, it is possible to include warehouse capacity constraints. The players may run a common warehouse whose size might be limited. For a contribution regarding warehouse capacity constraints, but without cooperation, compare, e.g., Love (1973). Let L denote the capacity limit of the warehouse:

$$\max -o + \sum_{i \in N} \pi_i z_i$$

s.t.

$$I_t = I_{t-1} + q_t - \sum_{i \in N} d_{it} z_i \qquad\qquad t = 1, \ldots, T$$

$$I_t \le L \qquad\qquad t = 1, \ldots, T$$

$$q_t \le M\, x_t \qquad\qquad t = 1, \ldots, T$$

$$o = \sum_{t=1}^{T}(c_t^p q_t + c_t^s x_t + c_t^h I_t)$$

$$I_t, q_t \geq 0 \qquad\qquad\qquad\qquad\qquad t = 1, \ldots, T$$
$$x_t \in \{0, 1\} \qquad\qquad\qquad\qquad\qquad t = 1, \ldots, T$$
$$z_i \in \{0, 1\} \qquad\qquad\qquad\qquad\qquad i \in N$$
$$o \geq 0.$$

Furthermore, we can extend the ELS setting to multilevel structures. Imagine, every player has implemented a multilevel order processing; e.g., via central and regional warehouses. On each level j, cooperation in-between those players is possible. For convenience, assume the case of a serial structure with a unique downstream successor $j + 1$ of stage j and J being the final stage. Compare Zangwill (1969) for an early work on serial lot sizing (but again without considering cooperation). This setting can be formulated as:

$$\max -o + \sum_{i \in N} \pi_i z_i$$

s.t.

$$I_{Jt} = I_{J,t-1} + q_{Jt} - \sum_{i \in N} d_{it} z_i \qquad\qquad\qquad t = 1, \ldots, T$$

$$I_{jt} = I_{j,t-1} + q_{jt} - q_{j+1,t} \qquad\qquad j = 1, \ldots, J-1, \ t = 1, \ldots, T$$

$$q_{jt} \leq M \, x_{jt} \qquad\qquad\qquad\qquad j = 1, \ldots, J, \ t = 1, \ldots, T$$

$$o = \sum_{j=1}^{J}\sum_{t=1}^{T}(c_{jt}^p q_{jt} + c_{jt}^s x_{jt} + c_{jt}^h I_{jt})$$

$$I_{jt}, q_{jt} \geq 0 \qquad\qquad\qquad\qquad j = 1, \ldots, J, \ t = 1, \ldots, T$$
$$x_{jt} \in \{0, 1\} \qquad\qquad\qquad\qquad j = 1, \ldots, J, \ t = 1, \ldots, T$$
$$z_i \in \{0, 1\} \qquad\qquad\qquad\qquad\qquad\qquad i \in N$$
$$o \geq 0.$$

We will study a more complex cooperative multilevel lot sizing problem in Chap. 8.

Chapter 6
A Lot Sizing Game with Uncertain Demand

In Chap. 5, we presented an economic lot sizing game where several players face dynamic demand that should be satisfied while making joint orders. Joint orders can reduce total costs because fixed ordering costs can be shared among the partners. In the following sections, we will introduce an extension of the before presented ELS game: In many practical situations, demand cannot be forecasted with certainty. Thus, a model that could handle uncertainties might be an interesting new feature for the ELS game.

6.1 The Underlying Problem

As in Drechsel and Kimms (2009), we take the economic lot sizing problem as discussed in Chap. 5 as a basis. The players' demand may be a source for uncertainty. Hence, assume that for each player i the demand in a certain period t is specified by an interval $[\underline{d}_{it}; \overline{d}_{it}]$ instead of a single value d_{it} as before. Consequently, total demand $d_t(N)$ in period t lies in the interval

$$[\underline{d}_t(N); \overline{d}_t(N)] = \left[\sum_{i \in N} \underline{d}_{it}(N); \sum_{i \in N} \overline{d}_{it}(N) \right].$$

From these interval bounds, the characteristic functions $\underline{c}^{IV}$ and $\overline{c}^{IV}$ can be derived by computing the optimal objective function values of the cooperative economic lot sizing problem (5.6)–(5.10) with the parameters $\underline{d}_t(S)$ and $\overline{d}_t(S)$ for $S \subseteq N$, respectively. By running the row generation procedure twice – once for $\underline{C}^{IV}(N, \underline{c}^{IV})$ and a second time for $\overline{C}^{IV}(N, \overline{c}^{IV})$ – it is possible to determine an element in the interval core.

For visualization, we introduce an example with $|N| = 3$ players and $T = 6$ periods. Table 6.1 displays the parameters for the interval ELS game – fixed and unit cost coefficients for ordering, unit cost coefficients for holding, and intervals for the uncertain demand. Applying the proposed row generation algorithm, we get the following interval core: $\pi(\underline{c}^{IV}) = (210, 510, 60)$ and $\pi(\overline{c}^{IV}) = (165, 615, 90)$.

J. Drechsel, *Cooperative Lot Sizing Games in Supply Chains*, Lecture Notes in Economics and Mathematical Systems 644, DOI 10.1007/978-3-642-13725-9_6,
© Springer-Verlag Berlin Heidelberg 2010

Table 6.1 Interval ELS game: parameters for the numerical example

t	1	2	3	4	5	6
$d_{1t} = [\underline{d}_{it}; \overline{d}_{it}]$	[10;15]	[10;15]	[10;15]	[10;15]	[10;15]	[10;15]
$d_{2t} = [\underline{d}_{it}; \overline{d}_{it}]$	[10;15]	[10;15]	[10;15]	[10;15]	[10;15]	[10;15]
$d_{3t} = [\underline{d}_{it}; \overline{d}_{it}]$	[10;15]	[10;15]	[10;15]	[10;15]	[10;15]	[10;15]
c_t^s	100	100	100	100	100	100
c_t^h	5	5	5	5	5	5
c_t^p	1	1	1	1	1	1

Table 6.2 Interval ELS game: characteristic function values for the numerical example

S	$\emptyset$	$\{1\}$	$\{2\}$	$\{3\}$	$\{1, 2\}$	$\{1, 3\}$	$\{2, 3\}$	$\{1, 2, 3\}$
$\underline{c}^{IV}(S)$	0	510	510	510	720	720	720	780
$\overline{c}^{IV}(S)$	0	615	615	615	780	780	780	870

6.2 Special Phenomena of Interval Cores

While using the classical core, one has to be aware that core allocations are not monotonic in general (see e.g., Young 1985). Drechsel and Kimms (2009) state that this effect might cause interpretation problems in the context of interval cores. See again the numerical example presented in Sect. 6.1. As the values for the characteristic function show (see Table 6.2), $\overline{c}^{IV}(S) \geq \underline{c}^{IV}(S)$ holds for this example. For an ELS game with non-negative cost data, this is always true because an increase in demand will not decrease the (total) costs (see Sect. 5.1.2). Thus, we would expect that an increasing demand of one player does not lead to a lower cost assignment; i.e., the principle of monotonicity holds also for the cost shares $(\pi_i(\overline{c}^{IV}) \geq \pi_i(\underline{c}^{IV}))$ because the characteristic functions $\underline{c}^{IV}$ and $\overline{c}^{IV}$ define the core cost allocations $\pi_i(\underline{c}^{IV})$ and $\pi_i(\overline{c}^{IV})$, respectively. This is not true in general, though. The numerical example presented before proves this: A lower cost share would be assigned to player 1 (165 instead of 210), although its demand increases from 10 to 15 for every period.

Young (1985) discusses monotonicity regarding allocation methods and proves that only the Shapley value is a unique symmetric and (strongly) monotonic allocation procedure. However, applying the Shapley value to interval-valued games and the interval core would bring another problem along: The Shapley value does not need to be in the core if the game is not concave – recall that the ELS game is not concave in general (see Sect. 5.1.2).

Hence, we might observe intervals $I_i = [\underline{L}_i; \overline{I}_i]$ with $\underline{L}_i > \overline{I}_i$. For our purposes, we accept this uncommon notation to keep the notation easy. However, $I_i = [\overline{I}_i; \underline{L}_i]$ is meant while interpreting such cases. This leads to the following intervals for our numerical example: $I_1 = [165; 210]$, $I_2 = [510; 615]$, and $I_3 = [60; 90]$.

The question remains, which concrete cost shares the players have to bear after the uncertainty has been resolved. The interval core could be interpreted as a promise to the players such that if $(I_1, \ldots, I_{|N|}) \in C^{IV}(N, c^{IV})$ is chosen in advance, an ex post (i.e., after demand has been realized) core cost allocation with

Table 6.3 Interval ELS game: realized demand for the numerical example

t	1	2	3	4	5	6
d_{1t}	15	15	15	15	15	15
d_{2t}	10	10	10	10	10	10
d_{3t}	15	15	15	15	15	15

$\pi_i \in I_i$ can be found. This holds because the two extreme scenarios (upper and lower bounds for the expected demands) have been considered to define the interval core. Therefore, the problem of finding a core element ex post seems to have the following form:

$$
C_{I_1,\ldots,I_{|N|}}(N, c) = \left\{ \pi \in \mathbb{R}^{|N|} \;\middle|\; \sum_{i \in N} \pi_i = c(N) \text{ and } \sum_{i \in S} \pi_i \le c(S) \right.
$$

$$
\left. \text{for all } S \subset N,\ S \neq \emptyset \text{ and } \pi_i \in I_i \right\}. \tag{6.1}
$$

This interpretation is meaningful if and only if the underlying (ex post) characteristic function c is monotone over the interval defined by c^{IV} and coincides with $\underline{c}^{IV}$ and $\overline{c}^{IV}$ at the interval borders. However, this interpretation does not hold, for example, in the case of the before presented numerical example: Assume now that the demand data displayed in Table 6.3 could be observed. According to this demand realizations, the individual costs amount to $c(\{1\}) = 615$, $c(\{2\}) = 510$, and $c(\{3\}) = 615$ (see Table 6.2). The costs for the grand coalition are $c(\{1, 2, 3\}) = 840$. The stability constraint for player 2 combined with the given interval I_2 in the definition (6.1) requires $\pi_2 = 510$. Due to the efficiency constraint, $\pi_1 + \pi_3 = c(N) - \pi_2 = 840 - 510 = 330$. This is not feasible because of the intervals $I_1 = [165; 210]$ (i.e., $\pi_1 \le 210$) and $I_3 = [60; 90]$ (i.e., $\pi_3 \le 90$). Hence, the interpretation of an arbitrary element from the interval core in the sense of cost limits that will be assigned to the players is not straightforward.

6.3 A New Definition of the Interval Core and Its Computation

Drechsel and Kimms (2009) suggest to adjust the row generation procedure to avoid the described interpretation problems. We need to compute interval borders $\pi_i(\underline{c}^{IV})$ and $\pi_i(\overline{c}^{IV})$ that assure $\pi_i(\underline{c}^{IV}) \le \pi_i(\overline{c}^{IV})$. This can be done by computing the values $\pi_i(\underline{c}^{IV})$ and $\pi_i(\overline{c}^{IV})$ simultaneously instead of computing upper and lower core allocations separately. Drechsel and Kimms (2009) develop a modification of the master problem:

$MP^{IV}(\underline{S},\overline{S})$:

$$\min \underline{w} + \overline{w} \tag{6.2}$$

s.t.

$$\sum_{i \in N} \pi_i(\underline{c}^{IV}) = \underline{c}^{IV}(N) \tag{6.3}$$

$$\sum_{i \in S} \pi_i(\underline{c}^{IV}) - \underline{w} \le \underline{c}^{IV}(S) \qquad\qquad S \in \underline{S} \tag{6.4}$$

$$\sum_{i \in N} \pi_i(\overline{c}^{IV}) = \overline{c}^{IV}(N) \tag{6.5}$$

$$\sum_{i \in S} \pi_i(\overline{c}^{IV}) - \overline{w} \le \overline{c}^{IV}(S) \qquad\qquad S \in \overline{S} \tag{6.6}$$

$$\pi_i(\underline{c}^{IV}) \le \pi_i(\overline{c}^{IV}) \qquad\qquad i \in N \tag{6.7}$$

$$\pi_i(\underline{c}^{IV}), \pi_i(\overline{c}^{IV}) \in \mathbb{R} \qquad\qquad i \in N \tag{6.8}$$

$$\underline{w}, \overline{w} \ge 0. \tag{6.9}$$

This master problem formulation contains efficiency and stability constraints ((6.3)–(6.6)) for the two extreme core allocations as defined in (3.18) and (3.19) (see p. 51). Constraints (6.7) assure that $\pi_i(\underline{c}^{IV})$ yields essentially a lower bound for the interval.

Due to the changed master problem, some modifications in the row generation procedure are necessary (see Sect. 3.2 for the original formulation of the procedure):

$Core^{IV}(MP^{IV}, SP)$:

1. Define small initial sets $\underline{S}$ and $\overline{S}$; e.g., $\underline{S} = \overline{S} = \{\{1\}, \{2\}, \dots, \{|N|\}\}$. Compute the individual total costs $\underline{c}^{IV}(S)$ and $\overline{c}^{IV}(S)$ for those coalitions $S \in \underline{S}$ and $S \in \overline{S}$ as well as the total costs $\underline{c}^{IV}(N)$ and $\overline{c}^{IV}(N)$ for the coalition N.
2. Solve the master problem $MP^{IV}(\underline{S},\overline{S})$ (6.2)–(6.9) optimally.
3. If $\underline{w} + \overline{w} > 0$, stop the algorithm because the instance has an empty interval core.
4. Otherwise:

   a. Find a coalition $\underline{S}' \notin \underline{S}$ ($\underline{S}' \ne \emptyset$) such that restriction $\sum_{i \in \underline{S}'} \pi_i(\underline{c}^{IV}) > \underline{c}^{IV}(\underline{S}')$ holds. Compute the total costs $\underline{c}^{IV}(\underline{S}')$ and update $\underline{S} = \underline{S} \cup \{\underline{S}'\}$ if such a coalition $\underline{S}'$ was found.
   b. Find a coalition $\overline{S}' \notin \overline{S}$ ($\overline{S}' \ne \emptyset$) such that restriction $\sum_{i \in \overline{S}'} \pi_i(\overline{c}^{IV}) > \overline{c}^{IV}(\overline{S}')$ holds. Compute the total costs $\overline{c}^{IV}(\overline{S}')$ and update $\overline{S} = \overline{S} \cup \{\overline{S}'\}$ if such a coalition $\overline{S}'$ was found.

5. If neither a coalition $\underline{S}'$ nor $\overline{S}'$ can be found, then stop the algorithm because the found allocations $\pi(\underline{c}^{IV})$ and $\pi(\overline{c}^{IV})$ are interval core elements.
6. Go to Step 2.

Since $\underline{S}'$ and $\overline{S}'$ may not be unique, Drechsel and Kimms (2009) suggest to formulate the subproblem in a way that it determines those coalitions $\underline{S}'$ and $\overline{S}'$, respectively, that violate the stability constraints most – thus, use the subproblem variant $\hat{SP}$ (see Sect. 5.2.1) in the form $\underline{\hat{SP}}(\pi(\underline{c}^{IV}))$ and $\overline{\hat{SP}}(\pi(\overline{c}^{IV}))$, respectively.

Recall the numerical example from Sect. 6.1. The presented algorithm yields the following allocation in the interval core: $\pi(\underline{c}^{IV}) = (210, 60, 510)$ and $\pi(\overline{c}^{IV}) = (210, 90, 570)$.

6.4 Computational Study for the Interval ELS Game

We tested the proposed row generation procedure for the interval core while using AMPL/CPLEX version 10.0.0 for the implementation and ran the tests on Intel Pentium hardware with 2.8 GHz and 504 MB RAM. As a basis, we used the randomly generated instances from Test-Bed 1 (see Sect. 5.3) to define the game $(N, \underline{c}^{IV})$. The game $(N, \overline{c}^{IV})$ has the same parameter values for the cost coefficients. However, the demand values were now chosen in the following way: Suppose $\underline{d}_{it}$ is the demand used in the corresponding game $(N, \underline{c}^{IV})$, then the demand $\overline{d}_{it}$ for the game $(N, \overline{c}^{IV})$ is a random integer drawn with uniform distribution $\overline{d}_{it} \in [\underline{d}_{it}; 20]$ to assure that $\underline{d}_{it} < \overline{d}_{it}$.

The average number of iterations as well as the minimum and maximum number of iterations over the 15 random instances for the row generation procedure are presented in Table 6.4. The table presents the results regarding number of iterations for the two subproblems $\underline{\hat{SP}}$ and $\overline{\hat{SP}}$ that are part of Step 4 of the row generation procedure. Recall that in the worst case, the number of iterations could have been in the order of $2^{|N|}$ which practically never happened. The computational time (measured in CPU-seconds) to determine an element in the interval core was less than a minute for all instances $|N| \leq 20$ and not more than five minutes for the larger instances.

Table 6.4 Interval ELS game: average results – average number of iterations (Minimum/Maximum number of iterations)

$|N|$	5	10	15	20	25	30
$\underline{\hat{SP}}$	4.87	20.47	146.27	406.73	1,044.60	1,043.53
	(4/7)	(14/40)	(25/293)	(110/1,036)	(165/2,364)	(30/3,946)
$\overline{\hat{SP}}$	4.73	21.27	146.27	403.80	1,039.80	1,054.53
	(4/6)	(12/60)	(16/293)	(110/1,037)	(165/2,360)	(30/3,941)

Chapter 7
A Capacitated Lot Sizing Game with Transshipments, Scarce Capacities, and Player-Dependent Cost Coefficients

In this chapter, we investigate a game with cooperative production based on the capacitated lot sizing problem where the available resources of the players may be used jointly. As in the chapters before, two topics will be discussed: Determining the optimal production plan for the grand coalition and allocating the total costs for such a production plan among the players.

7.1 Cooperative Production Situations

7.1.1 The Underlying Problem

In Chap. 5, a cooperative ordering situation with dynamic demand was presented. For practical situations, one should also take the shortness of resources for ordering or producing a product into account. Due to this, we now want to extend the economic lot sizing situation to a capacitated lot sizing problem (CLSP). The capacitated lot sizing problem is introduced by Billington et al. (1983) as an $\mathcal{NP}$-hard optimization problem.

We have one single decision maker i for whom we have to determine a production plan for T periods of time. In contrast to the assumptions of the ELS problem, several items can be produced. There is an index set K of the different products with the corresponding demand d_{ikt} for product k in period t that is faced by decision maker i. All products share a common resource that is available with R_{it} units per period. The parameter b_{ikt} gives the number of resource units needed to produce one product of type k in period t – we call it the capacity usage. Like for the ELS, we have different cost coefficients for setup (fixed cost per setup c_{ikt}^s, a setup is incurred whenever player i produces a positive quantity of product k in period t), warehousing (unit cost c_{ikt}^h if products k are stored over several periods until needed), and production (unit cost c_{ikt}^p). Apart from those parameters, the decision variables in this model are the production quantity q_{ikt} of product k in period t, the binary setup variable x_{ikt} that indicates in which period t a setup for which product k is needed, and the inventory level of product k at the end of period t I_{ikt}. I_{ik0} is a parameter and stands for the initial inventory of player i regarding item k at the beginning of the planning horizon ($t = 0$). The objective is to find a production plan with minimum costs while taking the trade off between setup and holding costs into account. The $CLSP_i$ for an individual decision maker i is a mixed-integer program:

$$\min \sum_{k \in K} \sum_{t=1}^{T} (c_{ikt}^s x_{ikt} + c_{ikt}^h I_{ikt} + c_{ikt}^p q_{ikt}) \tag{7.1}$$

s.t.

$$I_{ikt} = I_{ik,t-1} + q_{ikt} - d_{ikt} \qquad k \in K, t = 1,\ldots,T \tag{7.2}$$

$$\sum_{k \in K} q_{ikt} b_{ikt} \leq R_{it} \qquad t = 1,\ldots,T \tag{7.3}$$

$$q_{ikt} \leq M_{ikt} x_{ikt} \qquad k \in K, t = 1,\ldots,T \tag{7.4}$$

$$I_{ikt}, q_{ikt} \geq 0 \qquad k \in K, t = 1,\ldots,T \tag{7.5}$$

$$x_{ikt} \in \{0,1\} \qquad k \in K, t = 1,\ldots,T. \tag{7.6}$$

The objective function (7.1) minimizes total costs. The inventory balance restrictions are given by (7.2) – inventory level concerning product k at the end of period t results from starting inventory (at the beginning of period t) of product k plus the produced quantity of product k in period t minus the demand of this product k in period t. Constraints (7.3) assure that in every period t capacity R_{it} used for producing all products is not exceeded. Due to constraints (7.4), a setup has to be executed if production for product k in period t takes place. M_{ikt} is a parameter representing a sufficiently large number. For instance, it can be denoted as $M_{ikt} = \sum_{\tau=1}^{T} d_{ik\tau}$. (7.5) and (7.6) determine the domains of the decision variables.

Obviously, it is easily possible to extend the presented standard CLSP-model formulation (7.1)–(7.6) for additional aspects such as setup-times, backlogging, overtime, limited stocking capacity, and multi-level-product structures. For the sake of simplicity, we do not integrate these extensions into our model. However, the following discussion regarding the solution of the cooperative CLSP and cost allocation methods can be adapted straightforwardly to such extensions.

There exist many contributions in the literature regarding solution procedures and extensions for the CLSP and related problems. Recently, Buschkühl et al. (2010) and Quadt and Kuhn (2008) give actual reviews. Buschkühl et al. (2010) use the multilevel CLSP as basis for their survey and present modeling approaches plus algorithmic solution approaches from the last 40 years of international research. They disregard approaches concerning detailed sequencing and scheduling but point out that many practical problems are far from being solved satisfactorily. Quadt and Kuhn (2008) specifically concentrate on extensions for the CLSP; e.g., back-orders, setup carry-over, sequencing, and parallel machines. They present mathematical formulations and solution approaches for all the extensions. Drexl and Kimms (1997) review simultaneous lot sizing and scheduling. Mathematical models and further research opportunities are presented by them. Karimi et al. (2003) focus on modeling and solution approaches concerning single stage capacitated lot sizing problems.

7.1.2 The CLSP Game

Now, we extend the situation of the CLSP for a single decision maker to a cooperative setting. To the best of our knowledge, there is only one paper by Sambasivan and Yahya (2005) dealing with a related problem. They were motivated by a real-world problem coming from a company manufacturing steel rolled products. They develop a Lagrangean relaxation based heuristic to solve the problem but do not discuss the problem of cost allocation among the cooperating plants.

The idea behind the cooperative CLSP (or CLSP game), presented in Drechsel and Kimms (2010b), is that several decision makers (players) work together. That means a player cannot only produce items for its own demand but also to meet the demand of other players (obviously, only if the capacities are sufficient). N ($|N| \geq 2$) will be the index set of all players who are facing the same production situation. The incentive for cooperation is to reduce costs and avoid shortages because if the capacity of one player is insufficient to fulfill its demand, other players can produce the missing items or other players produce the items for lower costs if they have enough capacity. In fact, the players are exchanging resources in such a cooperation. A subset of players using their capacities jointly is called a coalition.

Apart from the assumptions for the $CLSP_i$, we now allow that the players pool their resources. Therefore, items need to be transported from the producing player to the player with the demand. Thus, we have an additional cost coefficient c_{ijkt}^t denoting the unit cost per item k shipped in period t from player i to player j. The transportation quantity of item k in period t from player i to player j is a_{ijkt}. The problem of finding an optimal production plan for a subcoalition $S \subset N$ or the grand coalition N can be formulated as follows:

$$c(S) = \min \sum_{i \in S} \sum_{k \in K} \sum_{t=1}^{T} \left(c_{ikt}^s x_{ikt} + c_{ikt}^h I_{ikt} + c_{ikt}^p q_{ikt} + \sum_{j \in S} c_{ijkt}^t a_{ijkt} \right) \quad (7.7)$$

s.t.

$$I_{ikt} = I_{ik,t-1} + q_{ikt} + \sum_{j \in S} a_{jikt} - d_{ikt} - \sum_{j \in S} a_{ijkt} \quad i \in S, k \in K, t = 1, \ldots, T \quad (7.8)$$

$$\sum_{k \in K} q_{ikt} b_{ikt} \leq R_{it} \quad i \in S, t = 1, \ldots, T \quad (7.9)$$

$$q_{ikt} \leq M_{kt} x_{ikt} \quad i \in S, k \in K, t = 1, \ldots, T \quad (7.10)$$

$$a_{ijkt} \geq 0 \quad i, j \in S, k \in K, t = 1, \ldots, T \quad (7.11)$$

$$I_{ikt},\, q_{ikt} \geq 0 \qquad\qquad\qquad\qquad i \in S,\, k \in K,\, t = 1,\dots,T \quad (7.12)$$

$$x_{ikt} \in \{0, 1\} \qquad\qquad\qquad\qquad i \in S,\, k \in K,\, t = 1,\dots,T. \quad (7.13)$$

The values of the decision variables q_{ikt} show the production quantity of each player (for itself and other players $j \in S$). There is no need to separate q in quantities produced for the player itself and other players (one might suggest to use a q_{ijkt} denoting the quantity of item k in period t produced by player i for player j). Having the optimal production plan for the cooperation, we can determine the quantities produced for other players with the help of the transportation quantity a_{ijkt}. A reason to do such a separation would be a different evaluation of quantities produced for oneself or for others. But for the sake of simplicity, we skip this assumption.

The objective function (7.7) still minimizes total costs, but compared to the $CLSP_i$ (see (7.1)), transshipment costs are now included. Similarly, we have to consider transshipment quantities in the inventory balances (7.8): The inventory level of player i and product k at the end of period t results from the starting inventory, the produced quantity, the quantity player i gets from other players $j \in S$, and minus the demand as well as the quantity transported from i to other players $j \in S$. The underlying assumption is that such transshipments happen within a cooperation only. The remaining constraints are taken from the standard CLSP (7.3)–(7.6). Again, M_{kt} is a sufficiently large parameter; e.g., $M_{kt} = \sum_{i \in N} \sum_{\tau=1}^{T} d_{ik\tau}$.

$c(S)$ denotes the characteristic function of the cooperative game (N, c). By construction, $c(\emptyset) = 0$ and $c(S) \geq 0$. To be well defined, let us assume $c(S) = \infty$ if the optimization problem has no feasible solution for the coalition S – this might happen if the capacities in the coalition S are not sufficient to satisfy the demand of the coalition S.

Note, transshipments from a player i to a player j need not be direct. Depending on the cost coefficient for transportation c_{ijkt}^t, it may happen that a product produced by player i in period t is transported from i via one or more other players before the final destination – i.e., fulfilling demand of player j in a period $t' \geq t$ – is reached. The product may even be stored by one or more players on its way to the final destination. If we make further triangle-inequality-like assumptions regarding transportation and holding costs, all transshipments are direct. However, the model presented above can handle arbitrary cost data.

Extensions, like limited transportation capacity, fixed transportation costs or the possibility for backlogging can easily be added to the model. It should be remarked that while including backlogging, one should add an equation to assure that a player can only transport quantities to another player if it produces enough or has enough inventory:

$$I_{ik,t-1} + q_{ikt} \geq \sum_{j \in S} a_{ijkt} \qquad\qquad i \in S,\, k \in K,\, t = 1,\dots,T.$$

Such restrictions avoid that a player's demand is satisfied by the backlog of another player.

An interesting interpretation of the lot sizing problem with transshipments is the lot sizing problem with substitutions. Axsäter (2003) studies a stochastic inventory situation with several warehouses. In-between those warehouses are lateral transshipments possible if a warehouse is running out of stock. Additionally, he considers another interpretation namely product substitution which means that a product running out of stock can be replaced by another product (e.g., one with higher quality) to avoid shortages. Lang and Domschke (2010) study product substitution in the context of capacitated and uncapacitated lot sizing problems as well. They focus on simple plant location based reformulations of the standard model.

A Numerical Example

We introduce a small numerical example to illustrate the above presented assumptions for the cooperative CLSP. Assume a game with three players ($N = \{1, 2, 3\}$), one product ($K = \{1\}$), and six periods of time ($T = 6$) (see Drechsel and Kimms 2010b). The cost coefficients are time-independent and given in Table 7.1. Each player i has $R_{it} = 40$ capacity units available in each period t. The capacity usage to produce one item is $b_{i1t} = 2$ for all players i and all periods t. Table 7.2 displays the demand that has to be met by every player. Without cooperation, each player has to produce the demand by its own production. The optimal production plans for this case are presented in Fig. 7.1. The circles represent the time periods for each player. The numbers within the circles display the production quantities p_{i1t}. Horizontal arcs mark that the products are kept on stock from one period to the next. The arc weights show the number of stored items. The total costs are $c(\{1\}) + c(\{2\}) + c(\{3\}) = 1,805$.

If the three players decide to cooperate in the grand coalition N, then an optimum solution might contain transshipments from one player to another. Figure 7.2 presents the optimal solution where transshipments are denoted by vertical arcs. The arc weights give the transportation quantity a_{ij1t}. Overall costs for the case

Table 7.1 CLSP game: cost coefficients for the numerical example

	$i = 1$	$i = 2$	$i = 3$	c^t_{ij1t}	$j = 1$	$j = 2$	$j = 3$
c^s_{i1t}	100	150	75	$i = 1$	0	10	10
c^p_{i1t}	0	0	0	$i = 2$	10	0	10
c^h_{i1t}	7	5	10	$i = 3$	10	10	0

Table 7.2 CLSP game: demand data for the numerical example

$t =$	1	2	3	4	5	6
d_{11t}	10	15	10	25	20	20
d_{21t}	10	15	15	10	20	5
d_{31t}	5	7	10	5	10	5

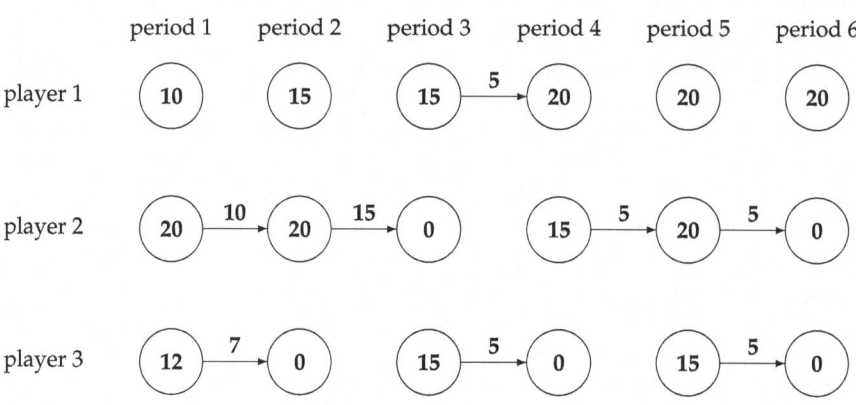

Fig. 7.1 CLSP game: an optimum solution without cooperation for the numerical example (Drechsel and Kimms 2010b)

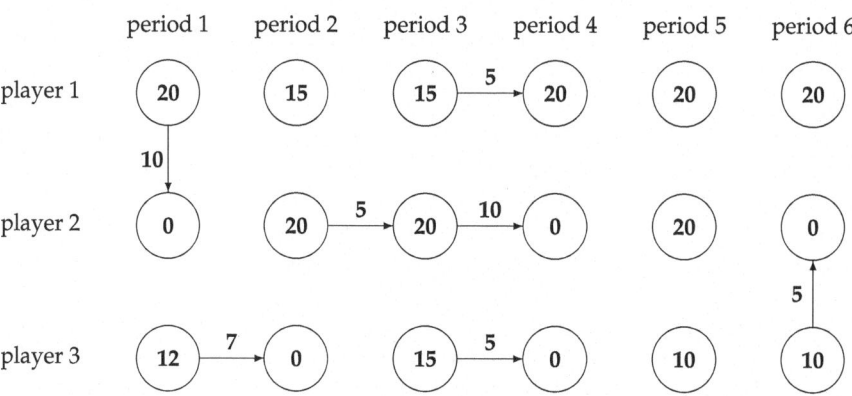

Fig. 7.2 CLSP game: an optimum solution with cooperation for the numerical example (Drechsel and Kimms 2010b)

Table 7.3 CLSP game: characteristic function values for the numerical example

S	$\{1\}$	$\{2\}$	$\{3\}$	$\{1,2\}$	$\{1,3\}$	$\{2,3\}$	$\{1,2,3\}$
$c(S)$	635	775	395	1,375	1,010	1,100	1,730

with cooperation are $c(\{1,2,3\}) = 1{,}730$. Obviously, cooperation leads to lower total costs.

We complete our example with the optimal objective function values for every possible coalition $S \subseteq N$ (see Table 7.3).

7.1.3 Properties of the CLSP Game

In this section, we investigate some important properties of the CLSP game: If the characteristic function c is *monotone*, (2.1) (see p. 11) must hold. It seems to be that with every new player coming in the coalition, total demand rises and, thus, total costs for producing the desired demand do not decrease (see the ELS game, Sect. 5.1.2). However, imagine the following case: One player i is joining an existing coalition S_1. This player has no demand over the whole time horizon but plenty of free capacity. It is possible that in the new coalition $S_2 = S_1 \cup \{i\}$ total costs can be reduced if production, setup and transportation costs of player i are smaller than in the coalition S_1. Consider a very simple example with two players ($|N| = 2$), one product ($|K| = 1$), one period ($T = 1$), a simple demand pattern $d_{111} = 1$, $d_{211} = \epsilon$, and the capacity pattern $R_{11} = R_{21} = 1 + \epsilon$ with $b_{111} = b_{211} = 1$. Assume that player 2 has lower production cost coefficients than player 1; i.e., $c_{211}^s < c_{111}^s$ and $(1 + \epsilon)c_{211}^p < c_{111}^p$. Furthermore, assume that the transportation cost coefficient follows $c_{2111}^t < c_{111}^p - (1 + \epsilon)c_{211}^p$. All other parameter values may have arbitrary non-negative values. The interpretation is that player 2 owns unused resources which can be cheaply provided to player 1. For the example, we have

$$c(\{1\}) = c_{111}^s + c_{111}^p \text{ and } c(\{2\}) > 0.$$

For the grand coalition, we get

$$c(\{1, 2\}) = c_{211}^s + (1 + \epsilon)c_{211}^p + c_{2111}^t < c_{211}^s + (1 + \epsilon)c_{211}^p + c_{111}^p - (1 + \epsilon)c_{211}^p$$
$$< c_{111}^s + c_{111}^p = c(\{1\}).$$

Hence, the characteristic function for the CLSP game is not monotone in general. Following the fact that monotone cooperative games only have non-negative cost assignments (see Drechsel and Kimms 2010c and Sect. 2.3.4, p. 24), there might be players who get a negative cost share which can be interpreted as that they are paid for participating in the coalition. In the before discussed example, the maximum cost share of player 1 would be $\pi_1 = c(\{1\})$. Caused by the efficiency constraint of the core ($\pi_1 + \pi_2 = c(\{1, 2\})$), this would lead to a negative cost share $\pi_2 = c(\{1, 2\}) - c(\{1\}) < 0$.

To demonstrate that there is an incentive to cooperate in such a cooperative CLSP situation with transshipments we have to prove *subadditivity* by (2.2) (see p. 12): Consider two disjoint and nonempty sets of players S_1 and S_2. We get exactly the same constraints of the above model for $c(S_1) + c(S_2)$ and $c(S_1 \cup S_2)$. That means the solutions (optimal production plans) for S_1 and S_2 together define a feasible solution of (7.7)–(7.13) for $S = S_1 \cup S_2$. Hence, the optimal value for $c(S_1 \cup S_2)$ cannot be greater than $c(S_1) + c(S_2)$ and the CLSP game is subadditive. In other words, if costs would not decrease while cooperating, solving $CLSP(S_1 \cup S_2)$ gives the same result as solving $CLSP(S_1)$ and $CLSP(S_2)$, separately, and add the objective function values. In this case, no transshipments between the coalitions S_1

and S_2 would take place. Therefore, we can just solve the problem for the grand coalition ($CLSP(N)$) and the variables a_{ijkt} show the effective coalition structure.

For choosing a procedure to compute stable cost shares, it is interesting to know whether the characteristic function c of a CLSP game is concave in general. *Concavity* is described by (2.3) (see p. 13). Having a look at the small example from Sect. 7.1.2, we can see that the CLSP game does not define a concave game in general:

$$c(\{1,2\}) - c(\{2\}) = 1{,}375 - 775 = 600$$
$$< 630 = 1{,}730 - 1{,}100 = c(\{1,2,3\}) - c(\{2,3\}).$$

7.2 Solving the Cooperative CLSP

This section will clarify how the optimization problem (7.7)–(7.13) can be solved if a subset $S \subseteq N$ of players decides to cooperate. Since the standard CLSP is an $\mathcal{NP}$-hard problem, the CLSP with transshipments as discussed in Sect. 7.1.2 is $\mathcal{NP}$-hard as well. Therefore, heuristics will be more appropriate to solve the problem than exact solution methods. Even if no transportation costs are considered, the remaining optimization problem is not an instance of the standard CLSP model. However, it can be interpreted as a CLSP variant with parallel machines – machines display the players in our cooperative setting. Nevertheless, in classical parallel machine lot sizing situations, the machines produce for a common demand. However, in our cooperative CLSP, each machine (player) faces its individual demand. There are few contributions to the topic of parallel machines in the literature (see e.g., Quadt and Kuhn 2008). Up to now, the variant including transportation costs is only studied by Sambasivan and Yahya (2005). The latter develop a heuristical solution approach but test only very small instances (up to 4 players, 6 periods of time, and 15 products). Summarized, our cooperative CLSP is a CLSP variant with parallel machines, machine-dependent demand, and transportation costs.

In the following two sections, we will develop two different heuristical approaches to solve the cooperative CLSP – a Lagrangean relaxation based heuristic and a fix-and-optimize approach.

7.2.1 A Lagrangean Relaxation Based Heuristic

We propose to use a Lagrangean relaxation of the capacity constraint (7.9). Approaches based on Lagrangean relaxation are already applied to other CLSP variants and the literature contributions show that it works successfully. Millar and Yang (1994) consider the CLSP variant with backorders and relax the setup constraints of their mathematical formulation. Based on the relaxation, they obtain two subproblems (one is a transportation problem) and develop two solution algorithms and valid lower bounds. A Lagrangean relaxation based heuristic for the multilevel

CLSP is studied by Tempelmeier and Derstroff (1996). They relax the capacity constraints and inventory balance equations and obtain single-item dynamic lot sizing subproblems of Wagner–Whitin type. Furthermore, they build up an effective heuristic. Grünert (1998) contributes substantial results regarding the multilevel sequence-dependent lot sizing and scheduling problem. He develops a Lagrangean tabu search algorithm to solve the problem efficiently. Özdamar and Barbarosoğlu (1999) contribute to the field of the multilevel CLSP as well. Furthermore, they include loading aspects in their model formulation and present two hybrid heuristics – one integrates simulated annealing and a genetic algorithm, whereas the second combines simulated annealing with a Lagrangean relaxation (relax the coupling constraint for assigning lots to facilities). The CLSP variant with linked lot sizes (setup carry-over into the next period is allowed) is studied by Sox and Gao (1999). For their Lagrangean decomposition heuristic, the capacity constraints and the constraints assuring that the setup for exactly one product is carried over into the next period are relaxed. Diaby et al. (1992) investigate the CLSP variant with multiple resources. This is different from multiple machines, because with multiple resources only one setup has to be carried out whenever a product is produced in a period independent from the number of different resources used. They focus on the solution of very-large-scale problems (5,000 products, 30 periods) with a Lagrangean relaxation based heuristic procedure (relaxing the constraints that incur the appropriate amounts of setup and processing time). Derstroff (1995) provides a comprehensive discussion regarding the multilevel CLSP with parallel machines (but without transshipments). He develops a Lagrangean relaxation procedure while relaxing capacity and inventory flow constraints. Last, Sambasivan and Yahya (2005) study a Lagrangean relaxation based heuristic for the cooperative CLSP.

Lower Bounds

The name "Lagrangean relaxation" is coined by Geoffrion (1974) who contributes substantially to this topic. Held et al. (1974) discuss the method of subgradient optimization in the context of Lagrangean relaxation. See also Fisher (1981) for a review regarding Lagrangean relaxation and a discussion of the generated bounds. We will build up our procedure on these basic ideas.

Sambasivan and Yahya (2005) propose a Lagrangean relaxation of the capacity restrictions (7.9) because the remaining problem is an uncapacitated lot sizing problem with transshipments. Let $\lambda_{it} \geq 0$ be the Lagrangean multipliers corresponding to restrictions (7.9). The relaxation results in the following model formulation:

$$
c^{LR}(N, \lambda) = \min \sum_{i \in N} \sum_{t=1}^{T} \left(\sum_{k \in K} \left(c_{ikt}^s x_{ikt} + c_{ikt}^h I_{ikt} + c_{ikt}^p q_{ikt} + \sum_{j \in N} c_{ijkt}^t a_{ijkt} \right) \right.
$$
$$
\left. + \lambda_{it} \left(\sum_{k \in K} b_{ikt} q_{ikt} - R_{it} \right) \right) \tag{7.14}
$$

s.t. (7.8), (7.10)–(7.13).

We set $\tilde{c}^p_{ikt} = c^p_{ikt} + \lambda_i b_{ikt}$ and reformulate the objective function (7.14):

$$\min \sum_{i \in N} \sum_{k \in K} \sum_{t=1}^{T} \left(c^s_{ikt} x_{ikt} + c^h_{ikt} I_{ikt} + \tilde{c}^p_{ikt} q_{ikt} + \sum_{j \in N} c^t_{ijkt} a_{ijkt} \right) - \underbrace{\sum_{i \in N} \sum_{t=1}^{T} \lambda_{it} R_{it}}_{*}.$$

The term (*) can be omitted for the purpose of optimization because it is constant. Thus, the objective function of the Lagrangean relaxation has the same structure as our original objective function (7.7) and, indeed, the resulting problem is an uncapacitated lot sizing problem with transshipments (contrary to the ELS). The production plans regarding the products do not depend on each other due to the relaxed capacity constraints. Hence, the remaining problem can be solved by computing $|K|$ single-product uncapacitated lot sizing problems with transshipments. Using standard software like CPLEX, for instance, and given values for λ_{it}, the problem can be formulated as

$$c^{LR}(N, \lambda) = -\sum_{i \in N} \sum_{t=1}^{T} \lambda_{it} R_{it} + \sum_{k \in K} v(k)$$

where $v(k)$ denotes the subproblem for product k:

$$v(k) = \min \sum_{i \in N} \sum_{t=1}^{T} \left(c^s_{ikt} x_{ikt} + c^h_{ikt} I_{ikt} + \tilde{c}^p_{ikt} q_{ikt} + \sum_{j \in N} c^t_{ijkt} a_{ijkt} \right)$$

s.t.

$$I_{ikt} = I_{ik,t-1} + q_{ikt} + \sum_{j \in N} a_{jikt} - d_{ikt} - \sum_{j \in N} a_{ijkt} \qquad i \in N, \, t = 1, \ldots, T$$

$$q_{ikt} \leq M_{kt} x_{ikt} \qquad\qquad\qquad i \in N, \, t = 1, \ldots, T$$
$$a_{ijkt} \geq 0 \qquad\qquad\qquad i, j \in N, \, t = 1, \ldots, T$$
$$I_{ikt}, q_{ikt} \geq 0 \qquad\qquad\qquad i \in N, \, t = 1, \ldots, T$$
$$x_{ikt} \in \{0, 1\} \qquad\qquad\qquad i \in N, \, t = 1, \ldots, T.$$

This uncapacitated single-item lot sizing problem with transshipments is still $\mathcal{NP}$-hard because it contains the $\mathcal{NP}$-hard uncapacitated facility location problem. Assume a small example with $T = 1$. Each player can be interpreted as a possible location for the production of product k, whereas the fixed costs for establishing a facility equal the setup costs. The customers' demand is defined by the demand of the players; i.e., the number of customers equals the number of players. The transportation cost coefficients can be converted into those for the facility location problem by solving a shortest path problem for each combination of players/customers.

Compare, for instance, Erlenkotter (1978) for detailed information regarding the uncapacitated facility location problem.

If we solve the above presented Lagrangean relaxation optimally for a given set of Lagrangean multipliers λ_{it}, we get a lower bound for the optimum objective function value $c(N)$ defined by (7.7); i.e., $c^{LR}(N, \lambda) \leq c(N)$. Depending on the predefined Lagrangean multipliers, this lower bound can be closer or more distant to the original objective function value $c(N)$. We propose an iterative procedure using subgradient optimization to find a good set of Lagrangean multipliers.

For the subgradient optimization, we define the unused capacity of player i in period t after a certain iteration in which the Lagrangean multipliers λ'_{it} were used:

$$\Delta_{it} = R_{it} - \sum_{k \in K} b_{ikt} q_{ikt}.$$

Δ_{it} is negative if the capacity constraint is violated. We start in iteration 0 with $\lambda_{it} = 0$. For the next iteration, the Lagrangean multipliers can be computed as:

$$\lambda_{it} = \max \left\{ 0, \lambda'_{it} - \delta \frac{(UB^* - LB^*) \Delta_{it}}{\sum_{i' \in N} \sum_{t'=1}^{T} \Delta_{i't'}^2} \right\}.$$

UB^* and LB^* denote the best known upper and lower bound until this iteration, respectively. δ is a control parameter that is initially set to $\delta = 2$ and divided by two whenever the best lower and the best upper bound has not changed within the last five iterations. Our iteration procedure continues as long as $\delta > 0.00001$ holds.

Upper Bounds

After solving the Lagrangean relaxation to get a lower bound, we can use those solutions for the relaxed problem to generate feasible solutions for the original cooperative CLSP problem serving as upper bounds. Thus, within each iteration of the Lagrangean relaxation procedure, we derive a heuristic solution. Sambasivan and Yahya (2005) propose a so-called lot shifting-splitting-merging routine to remove capacity violations – shift production quantities to plants where the demand occurs and production quantities within a plant are shifted across periods. Apart from this idea, several other ideas can be tested (see also Drechsel and Kimms 2010b):

h1: If there are setup variables x_{ikt} with a value equal to 1 in the solution of the Lagrangean relaxation, fix those variables. Afterwards, solve the model (7.7)–(7.13) which provides an upper bound of $c(N)$ – use the fixed setup variables and let the other decision variables be free, so they will be determined during this optimization. A feasible solution is assured if the cooperative CLSP has one.

h2: All setup variables x_{ikt} are fixed and then the remaining linear program (7.7)–(7.11) is solved. Possibly, the solution is not feasible because of the capacity restrictions. However, we may find feasible solutions in some iterations.

h3: Use the variant h1 but proceed it only in every 50th iteration of the subgradient optimization to reduce computational time.

h4: Fix all setup variables to one and solve the cooperative CLSP. In a post-processing step, we reduce the objective function value by the setup cost for those combinations of periods, players, and products where no production is planned. Obviously, this upper bound must be computed only once.

h5: Solve the LP relaxation of the cooperative CLSP. During a post-processing step, we "correct" the objective function value where $x_{ikt} \leq 1$ with the full setup costs. Again, this variant provides only one upper bound.

h6: As in variant h1, fix those setup variables whose value is equal to 1 after solving the Lagrangean relaxation. Solve the LP relaxation of the cooperative CLSP with the fixed setup variables – the other decision variables remain free and will be determined during the optimization. During a post-processing step, we set all non-integer x_{ikt} variables with $x_{ikt} > 0$ to one and compute the objective function value again (therefor no optimization problem need to be solved). This fixing procedure will always yield a feasible solution if the cooperative CLSP has one.

To all variants, we add a further post-processing step to eliminate needless setup costs c_{ikt}^s where the setup variable was fixed to one from the Lagrangean relaxation but the optimal solution of the remaining problem reveals no production quantity ($q_{ikt} = 0$). Standard software (e.g., CPLEX) can be used to solve the simplified cooperative CLSP which is of a smaller size than the original problem due to fixing some decision variables.

First tests with the variant h2 showed that in no iteration of the subgradient optimization a feasible solution could be found. Mostly, there are not enough setup variables fixed to one to assure feasible production. Due to the constant upper bound in variants h4 and h5, the lower bound generated in the subgradient optimization is not that good as in variants with changing upper bounds. Therefore, we skip these three variants for the computational study and inspect only h1, h3, and h6.

7.2.2 A Fix-and-Optimize Heuristic

Apart from such Lagrangean relaxation based heuristics, other approaches for solving CLSP variants are also presented in the literature. Recently, a fix-and-optimize procedure for the multilevel CLSP with positive lead times is introduced by Helber and Sahling (2010). Their algorithm brings substantial progress because it outperforms other approaches – like the Lagrangean decomposition method of Tempelmeier and Derstroff (1996) and the time decomposition approach of Stadtler (2003). Due to the promising results presented by Sahling et al. (2009) for a very complex CLSP variant (the multilevel CLSP with multiperiod setup carry-over), we modified this fix-and-optimize idea for the cooperative CLSP. Starting point is the following slightly modified model formulation of (7.7)–(7.13):

$$c^+(S) = \min \sum_{i \in S} \sum_{k \in K} \sum_{t=1}^{T} \left(c^s_{ikt} x_{ikt} + c^h_{ikt} I_{ikt} + c^p_{ikt} q_{ikt} + \sum_{j \in S} c^t_{ijkt} a_{ijkt} \right)$$

$$+ M^{inv} \sum_{i \in S} \sum_{k \in K} I^+_{ik0} \qquad (7.15)$$

s.t.

$$I_{ik1} = I^+_{ik0} + I_{ik0} + q_{ik1} + \sum_{j \in S} a_{jik1} - d_{ik1} - \sum_{j \in S} a_{ijk1} \quad i \in S, \, k \in K \quad (7.16)$$

$$I_{ikt} = I_{ik,t-1} + q_{ikt} + \sum_{j \in S} a_{jikt} - d_{ikt} - \sum_{j \in S} a_{ijkt} \quad i \in S, \, k \in K, \, t = 2, \ldots, T$$

$$(7.17)$$

$$\sum_{k \in K} b_{ikt} q_{ikt} \leq R_{it} \qquad\qquad i \in S, \, t = 1, \ldots, T \qquad (7.18)$$

$$q_{ikt} \leq M_{kt} x_{ikt} \qquad\qquad i \in S, \, k \in K, \, t = 1, \ldots, T \qquad (7.19)$$

$$x_{ikt} = \chi_{ikt} \qquad\qquad (i, k, t) \in X^{fix} \qquad (7.20)$$

$$a_{ijkt} \geq 0 \qquad\qquad i, j \in S, \, k \in K, \, t = 1, \ldots, T \qquad (7.21)$$

$$I^+_{ik0} \geq 0 \qquad\qquad i \in S, \, k \in K \qquad (7.22)$$

$$I_{ikt}, \, q_{ikt} \geq 0 \qquad\qquad i \in S, \, k \in K, \, t = 1, \ldots, T \qquad (7.23)$$

$$x_{ikt} \in \{0, 1\} \qquad\qquad i \in S, \, k \in K, \, t = 1, \ldots, T. \qquad (7.24)$$

We introduce a new decision variable I^+_{ik0} which specifies the quantity of item k allocated to player i to guarantee a feasible solution; i.e., the quantity to fulfill demand in addition to the production quantities and the available initial inventories – see (7.16) and (7.22). Thus, it is some kind of artificial variable to avoid problems of infeasibility that might occur if some of the other variables are fixed during the procedure of the heuristic. The set X^{fix} specifies the (sub-)set of those x_{ikt} variables who shall be fixed. The parameters $\chi_{ikt} \in \{0, 1\}$ define the fixed value. Restrictions (7.20) assure this procedure. M^{inv} is a large number which can be interpreted as a penalty to the objective function value (7.15) to assure $I^+_{ik0} = 0$ as long as the original cooperative CLSP, or the problem derived from that by fixing some variables, has a feasible solution. Note, $c^+(S) = c(S)$ if $I^+_{ik0} = 0$ for all i and k and $X^{fix} = \emptyset$. For all other cases, $c^+(S)$ specifies an upper bound for $c(S)$ – no matter how X^{fix} and the values χ_{ikt} are defined.

The General Procedure

Let us now describe the basic steps of the iterative fix-and-optimize heuristic that will be executed until a local optimum is found.

0. Initializing:
- $X^{fix} = \{(i,k,t)|i \in N; k \in K; t = 1,\ldots,T\}$.
- $\chi_{ikt} = 1$ for all $(i,k,t) \in X^{fix}$.
- Solve (7.15)–(7.24) optimally (with standard software like CPLEX).
- Add the post-processing to obtain $c^+(S)$.
- $UB = c^+(N)$.
- If $\sum_{i \in S} \sum_{k \in K} I^+_{ik0} > 0$, then stop because the instance has no feasible solution.

1. Start of an iteration: $LocOpt = true$
2. Fix-and-optimize:

   a. • Define X^{fix}.
      • Compute $c^+(N)$; i.e., solve (7.15)–(7.24) optimally (with standard software like CPLEX).
      • Perform post-processing.
      • If $c^+(N) < UB$, then update $UB = c^+(N)$, $LocOpt = false$, and $\chi_{ikt} = x_{ikt}$ for all $(i,k,t) \notin X^{fix}$.

   b. Repeat (2a) until all alternatives to define X^{fix} are investigated.

3. End of an iteration: If $LocOpt = false$, then repeat from Step 1.

The mentioned post-processing procedure is done similar to the method described for the Lagrangean relaxation based upper bounds (see Sect. 7.2.1): For all $(i,k,t) \in X^{fix}$, we test whether $x_{ikt} = 1$ and $q_{ikt} = 0$ is true. We subtract c^s_{ikt} from the objective function value for each such case. Note that no χ_{ikt} values are modified during this post-processing.

One important question is still open: How should X^{fix} be defined? Let $X^{all} = \{(i,k,t)|i \in N; k \in K; t = 1,\ldots,T\}$ be the set of all index tuples under consideration. Furthermore, let $X^{opt} = X^{all} \backslash X^{fix}$ denote the set of variables (index tuples) which are optimized. Hence, $X^{fix} = X^{all} \backslash X^{opt}$. We suggest to use all alternatives that result from the following decomposition principles:

- Player based decomposition:
  For each player i^* define $X^{opt} = \{(i,k,t) \in X^{all}|i = i^*\}$.
- Product based decomposition:
  For each product k^* define $X^{opt} = \{(i,k,t) \in X^{all}|k = k^*\}$.
- Time based decomposition:
  For the given parameters ΔT and $\overline{T}$ a time window within the interval $[1;T]$ can be specified by $W(w) = [w \cdot \Delta T + 1; \min\{w \cdot \Delta T + \overline{T}; T\}]$ with a multiplier $w \in \{0,\ldots,\max\{0; \lceil T - \overline{T}/\Delta T \rceil\}\}$. For each multiplier w define $X^{opt} = \{(i,k,t) \in X^{all}|t \in W(w)\}$.

That means, for an example with $|N| = 10$ players, $|K| = 3$ products, and $T = 6$ periods, we get ten player-based alternatives, three product-based alternatives and for $\Delta T = 2$ and $\overline{T} = 4$ two time-based alternatives. Hence, Step 2 of the fix-and-optimize procedure would be repeated 15 times per iteration.

The remarkable idea behind this fix-and-optimize approach is that the subproblem (including several fixed variables) has a much smaller number of variables to be optimized than the original cooperative CLSP. Although the complexity status of the problem stays the same, the remaining optimization problems could be handled by standard software.

A related concept is the so called relax-and-fix heuristic (see Stadtler 2003 and Pochet and Wolsey 2006, p. 109). In contrast to the approach of Helber and Sahling (2010), using relax-and-fix requires three groups of binary variables – the first contains fixed variables, the second those that are optimized, and the third includes variables with relaxed integrality constraints.

7.3 Computing Core Cost Allocations for the CLSP Game

After being able to solve the optimization problem behind the characteristic function of the CLSP game, we now tackle the problem of finding a cost allocation for the total costs $c(N)$ which meets specific properties, like stability and fairness. Goemans and Skutella (2004) show that determining whether or not a given allocation is in the core is $\mathcal{NP}$-complete for the unconstrained facility location problem – the same holds for testing whether or not the core is empty. The insights from Sect. 7.2.1 regarding the relation between the uncapacitated single-item lot sizing problem with transshipments and the uncapacitated facility location problem, hence, reveal the complexity status of the core cost allocation problem for the CLSP game.

7.3.1 The Row Generation Procedure

For finding a cost allocation in the core (which fulfills the request for stability), we use the row generation algorithm introduced in Drechsel and Kimms (2010a) which is already introduced in Sect. 3.2 and applied to the ELS game (see Sect. 5.2). The master problem (a linear program) is solved with standard software and provides a cost allocation π. A problem specific subproblem (mixed-integer) is solved with standard software as well and reveals whether or not this cost allocation π is in the core. In the latter case, the subproblem generates an additional constraint for the master problem. The procedure is repeated until the cost allocation π is in the core. Apart from finding a core cost allocation, the algorithm is able to detect whether or not the instance has an empty core.

For the master problem formulation, compare (3.1)–(3.5) in Sect. 3. The problem specific subproblem for the CLSP game can be formulated as follows. We refer to the subproblem version $\hat{SP}(\pi)$ (accordingly, the objective function maximizes the violation of the core defining stability constraint, see Sect. 5.2.1):

$\hat{SP}(\pi)$:

$$\max o + \sum_{i \in N} \pi_i z_i \tag{7.25}$$

s.t.

$$o = \sum_{i \in N} \sum_{k \in K} \sum_{t=1}^{T} \left(c_{ikt}^{s} x_{ikt} + c_{ikt}^{h} I_{ikt} + c_{ikt}^{p} q_{ikt} + \sum_{j \in N} c_{ijkt}^{t} a_{ijkt} \right) \qquad (7.26)$$

$$I_{ikt} = I_{ik,t-1} + q_{ikt} + \sum_{j \in N} a_{jikt} - d_{ikt} z_i - \sum_{j \in N} a_{ijkt}$$
$$i \in N, \, k \in K, \, t = 1, \dots, T \qquad (7.27)$$

$$\sum_{k \in K} b_{ikt} q_{ikt} \leq R_{it} \qquad\qquad i \in N, \, t = 1, \dots, T \qquad (7.28)$$

$$q_{ikt} \leq M_{kt} x_{ikt} \qquad\qquad i \in N, \, k \in K, \, t = 1, \dots, T \qquad (7.29)$$
$$x_{ikt} \leq z_i \qquad\qquad i \in N, \, k \in K, \, t = 1, \dots, T \qquad (7.30)$$
$$a_{ijkt} \leq M_{kt} z_i \qquad\qquad i, j \in N, \, k \in K, \, t = 1, \dots, T \qquad (7.31)$$
$$a_{ijkt} \geq 0 \qquad\qquad i, j \in N, \, k \in K, \, t = 1, \dots, T \qquad (7.32)$$
$$I_{ikt}, \, q_{ikt} \geq 0 \qquad\qquad i \in N, \, k \in K, \, t = 1, \dots, T \qquad (7.33)$$
$$x_{ikt} \in \{0, 1\} \qquad\qquad i \in N, \, k \in K, \, t = 1, \dots, T \qquad (7.34)$$
$$o \geq 0 \qquad\qquad (7.35)$$
$$z_i \in \{0, 1\} \qquad\qquad i \in N. \qquad (7.36)$$

For the decision variables a_{ijkt}, I_{ikt}, q_{ikt}, and x_{ikt} and the constraints (7.28) and (7.29), compare the model for the CLSP game (7.7)–(7.13). The decision variables o and z_i have the same interpretation as in the subproblem formulation for the ELS game (see Sect. 5.2.1) – o indicates the characteristic function value for the coalition S' and z_i calls the players belonging to S' ($z_i = 1$ means $i \in S'$ and $z_i = 0$ otherwise). Accordingly, the inventory balances (7.27) take into account which players belong to the subcoalition and consider only their demand. Constraints (7.30) assure that if a player i sets up for production for any product k in any period t ($x_{ikt} = 1$), this player has to be in the coalition ($z_i = 1$). In combination with (7.29), production ($q_{ikt} > 0$) outside the coalition is not possible. The same issue is applied to transportation events; i.e., only players in the coalition are allowed to ship items to others (in the coalition) – it is assured by (7.31). If the triangle inequality holds for the transportation costs c_{ijkt}^{t}, then there is no need for constraints (7.31). Otherwise, it might happen to transport products produced by player i for player j via another player l without player l being in the coalition. Therefore, we have to assure that all players who are included in such transports (and not in the production process) belong to the coalition. Otherwise, no transports via these players are allowed.

The whole row generation algorithm proceeds in the following steps (see Drechsel and Kimms 2010b):

$\hat{C}ore(MP, \hat{SP})$:

1. Define a small initial set S; e.g., $S = \{\{1\}, \{2\}, \ldots, \{|N|\}\}$. Compute the individual total costs $c(S)$ for those coalitions $S \in S$ and the total costs $c(N)$ for the coalition N.
2. Solve the linear program $MP(S)$ (see (3.1)–(3.5)) optimally.
3. If $w > 0$, stop the algorithm because the instance has an empty core.
4. Otherwise, solve $\hat{SP}(\pi)$ (see (7.25)–(7.36)) optimally.
5. If $\hat{SP}(\pi)$ has a non-positive optimum objective function value, then stop the algorithm because the found allocation is in the core.
6. Otherwise, set $c(S') = o$, update $S = S \cup \{S'\}$ and go to Step 2.

A Numerical Example

Consider again the example presented in Sect. 7.1.2. The before described row generation procedure is used to allocate the total costs of the grand coalition $c(N) = 1{,}730$ among the three players.

Iteration 1: The individual total costs and those for the grand coalition were already determined – $c(\{1\}) = 635$, $c(\{2\}) = 775$, $c(\{3\}) = 395$, and $c(\{1, 2, 3\}) = 1{,}730$. We solve the master problem $MP(\{1\}, \{2\}, \{3\})$ optimally:

$$\min w$$

s.t.

$$\pi_1 + \pi_2 + \pi_3 = 1{,}730$$
$$\pi_1 - w \leq 635$$
$$\pi_2 - w \leq 775$$
$$\pi_3 - w \leq 395$$
$$w \geq 0.$$

The optimum solution is $w = 0$ and $\pi = (635, 775, 320)$. Therefore, the core is not empty and the procedure can be continued with checking for a coalition violating not considered stability constraints. Solving the subproblem optimally results in $z_1 = z_2 = 1$ and $z_3 = 0$. That means that players 1 and 2 have an incentive not to work in the grand coalition ($\pi_1 + \pi_2 = 1{,}410 > 1{,}375 = c(\{1, 2\})$).

Iteration 2: We add the found coalition to S and solve $MP(\{1\}, \{2\}, \{3\}, \{1, 2\})$ optimally which means to solve the master problem from Iteration 1 with the additional stability constraint

$$\pi_1 + \pi_2 - w \leq 1{,}375.$$

The optimum solution is $w = 0$ with $\pi = (600, 775, 355)$. We call the subproblem that gives the result $z_1 = 0$, $z_2 = z_3 = 1$ because $\pi_2 + \pi_3 = 1{,}130 > 1{,}100 = c(\{2, 3\})$.

Iteration 3: Adding a further constraint to the master problem, we now solve $MP(\{1\}, \{2\}, \{3\}, \{1, 2\}, \{2, 3\})$ optimally which means to solve the master problem from Iteration 2 with the additional constraint

$$\pi_2 + \pi_3 - w \leq 1{,}100.$$

The optimum solution is $w = 0$ and $\pi = (630, 745, 355)$. Calling the subproblem reveals that this solution is in the core because no solution with a positive objective function value can be found. Hence, the algorithm terminates and the last cost allocation is in the core.

7.3.2 Computing the Subcoalition-Perfect Core

We already described that the CLSP game is not monotone in general (see Sect. 7.1.3). This is no problem for the presented row generation procedure. Nevertheless, computing an element in the subcoalition-perfect core seems to be more appropriate for such instances following the aforementioned argumentation (see Sect. 2.3.7). For the computation, we simply have to use the master problem $MP^+(\mathcal{S})$ (see Sect. 3.5) and the subproblem (7.25)–(7.36) presented in Sect. 7.3.1.

It should be emphasized once again that without the equivalence of the subcoalition-perfect core and the set of non-negative core allocations (see p. 33), the adaptation of the row generation algorithm would be more complicated. In the master problem, there would be no non-negativity conditions for the π_i values and Drechsel and Kimms (2010c) suggest to formulate the subproblem as follows:

$$\max -o + \sum_{i \in N} \pi_i z_i \tag{7.37}$$

s.t. (7.26), (7.28), (7.29), (7.32)–(7.36), and

$$I_{ikt} = I_{ik,t-1} + q_{ikt} + \sum_{j \in N} a_{jikt} - d_{ikt}z'_i - \sum_{j \in N} a_{ijkt} \qquad i \in N,\, k \in K,\, t = 1, \ldots, T \tag{7.38}$$

$$x_{ikt} \leq z'_i \qquad i \in N,\, k \in K,\, t = 1, \ldots, T \tag{7.39}$$

$$a_{ijkt} \leq M_{kt}z'_i \qquad i, j \in N,\, k \in K,\, t = 1, \ldots, T \tag{7.40}$$

$$z_i \leq z'_i \qquad i \in N \tag{7.41}$$

$$z'_i \in \{0, 1\} \qquad i \in N. \tag{7.42}$$

One would need two sets of binary variables – the set of variable z_i indicates membership in coalition S_1 and the variables z'_i identify members of the supercoalition S_2 for the subcoalition-perfect core defining constraint $\sum_{i \in S_1} \pi_i \leq c(S_2)$ with $S_1 \subseteq S_2 \subseteq N$. Constraints (7.41) assure $S_1 \subseteq S_2$. This subproblem formulation requires more run-time to find an element in the subcoalition-perfect core

Table 7.4 CLSP game: cost coefficients for the numerical example subcoalition-perfect core

	$i = 1$	$i = 2$	$i = 3$	c^t_{ij1t}	$j = 1$	$j = 2$	$j = 3$
c^s_{i1t}	100	150	75	$i = 1$	0	10	10
c^p_{i1t}	0	0	0	$i = 2$	10	0	10
c^h_{i1t}	7	5	10	$i = 3$	1	1	0

Table 7.5 CLSP game: demand data for the numerical example subcoalition-perfect core

$t =$	1	2	3	4	5	6
d_{11t}	10	15	10	25	20	20
d_{21t}	10	15	15	10	20	5
d_{31t}	1	0	0	0	0	0

Table 7.6 CLSP game: characteristic function values for the numerical example subcoalition-perfect core

S	$\{1\}$	$\{2\}$	$\{3\}$	$\{1,2\}$	$\{1,3\}$	$\{2,3\}$	$\{1,2,3\}$
$c(S)$	635	775	75	1,375	585	500	1,031

than the formulation with non-negative π_i values. Therefore, the equivalence of the subcoalition-perfect core and the set of non-negative core allocations does not only bring theoretical insight, but reduces computational effort for computing a subcoalition-perfect core allocation.

A Numerical Example

Let us show the effect of the subcoalition-perfect core by means of a small numerical example. The instance has three players ($N = \{1, 2, 3\}$), one product ($K = \{1\}$), and six periods of time ($T = 6$). The cost coefficients are time-independent and given in Table 7.4. Each player i has $R_{it} = 40$ capacity units available in each period t. The capacity usage to produce one item is $b_{i1t} = 2$ for all players i and all periods t. Note that we change the demand data and the transportation cost coefficients for player 3 to assure that the instance is non-monotone (see Table 7.5). Table 7.6 displays the characteristic function values. Due to, e.g., $c(\{1, 2, 3\}) < c(\{1, 2\})$, the instance is non-monotone.

Using the row generation procedure to compute an element in the core, the result is $\pi = (600, 775, -344)$ after two iterations. Hence, player 2 would prefer the coalition $S = \{2, 3\}$ over the grand coalition because he hopes to get a cost share that is less than $c(\{2, 3\}) = 500$. But when computing an element in the subcoalition-perfect core, the result is a non-negative cost allocation $\pi = (585, 446, 0)$ after two iterations which assures that the share of player i is equal or smaller than in any sub-coalition where player i participates. Additionally, using the alternative subproblem formulation (7.37)–(7.42) reveals the cost allocation $\pi = (531, 500, 0)$ after three iterations.

If we want to allow that player 3 should get some kind of reward for providing its capacity, another core variant (e.g., the minmax core) should be used.

7.3.3 Computing the Minmax Core

Another very promising core variant is introduced in Sect. 2.3.6 – the minmax core. Following a minmax principle, every player should participate equally on the relative benefit earned by the coalitions. To proceed the row generation algorithm, we have to use the master problem formulation (3.8)–(3.12) (see Sect. 3.5) and the subproblem (7.25)–(7.36) with some slight changes in the objective function and a new restriction:

$$\max -\eta \, o + \sum_{i \in N} \pi_i z_i \tag{7.43}$$

s.t. (7.26)–(7.36)

$$\sum_{i \in N} z_i \leq |N| - 1. \tag{7.44}$$

η needs to be included in the objective function (7.43). Furthermore, it might happen that the subproblem finds the grand coalition as the coalition violating the stability constraint most – caused by the efficiency restriction, the equation $o = \sum_{i \in N} \pi_i z_i$ must hold. Hence, (7.44) assures that the subproblems find only subcoalitions $S \subset N$.

If the last η value is equal to or smaller than one, this minmax core element is in the core; i.e., this allocation is stable.

A Numerical Example

Recall the instance used in Sect. 7.1.2. The row generation procedure reveals the following cost allocation in the minmax core after three iterations: $\pi = (633.17, 722.911, 373.919)$ with $\eta = 0.997118$. This result can be interpreted as follows: No single player and no subcoalition $S \subset N$ gets a higher cost share than 99.7% of its standalone costs.

7.4 Computational Study for the CLSP Game

The proposed procedures to solve the cooperative CLSP and the corresponding CLSP game (the cost allocation problem) were implemented in AMPL/CPLEX version 10.0.0. Most of the results presented in this section are taken from Drechsel and Kimms (2010b).

Randomly generated instances with $T = 6$ were used for the tests where the integral parameter values were drawn from the following intervals: $c_{ikt}^s \in [0; 200]$, $c_{ikt}^p \in [0; 15]$, $c_{ikt}^h \in [0; 10]$, $d_{ikt} \in [0; d^{\max}]$ with $d^{\max} = 20$, and $b_{ikt} = b_{ik} \in [1; 5]$. The definition for the interval of R_{it} is a little bit more complicated because we want to assure that every player has enough capacity over the planning horizon to produce at least its own demand: $R_{it} \in [\max\{0; \sum_{k \in K} \sum_{t' \in \{1,...,t\}} b_{ikt'} d_{ikt'} - \sum_{t' \in \{1,...,t-1\}} R_{it'}\}; \sum_{k \in K} b_{ikt} d^{\max}]$. If we would not define the instances in this way, there might be coalitions that are not feasible (due to the lack of capacity). Such a setting would lead to a special chapter in cooperative game theory called games with restricted cooperation which will be studied in more detail in Chap. 8.

For the heuristics solving the cooperative CLSP, we systematically varied three parameters: the number of players ($|N| \in \{3, 4, 5, 10, 15, 20, 25, 50, 100\}$), the number of products ($|K| \in \{1, 3, 15\}$), and the range of transportation costs. The random transportation cost coefficients were drawn from the following intervals: $c_{ijkt}^t \in [\underline{c}_{ijkt}^t; \overline{c}_{ijkt}^t]$ where $[\underline{c}_{ijkt}^t; \overline{c}_{ijkt}^t] \in \{[0; 0], [0; 5], [0; 15], [0; 50], [5000; 5000]\}$. Fifteen random instances were generated for each parameter combination. To compare the proposed heuristics later on, we solved benchmark instances with AMPL/CPLEX 10.0.0 optimally. Table 7.7 displays the average computation time for finding an optimum solution for instances with $c_{ijkt}^t \in [0; 15]$. The results show that commercial software packages can handle rather large instances.

For computing core cost allocations, we only used transportation cost coefficients $c_{ijkt}^t \in [0; 15]$. The number of players was systematically varied, $|N| \in \{3, 5, 10, 15\}$, in combination with $|K| = 3$ products. For the small coalition with $|N| = 3$, we additionally tested $|K| = 1$. 15 instances per parameter combination were tested.

7.4.1 Computational Study: Lagrangean Relaxation Based Heuristic

Sambasivan and Yahya (2005) present tests regarding instances with three to four players, three to six periods, and three to 15 products using their Lagrangean

Table 7.7 CLSP game: average CPU time optimal solutions with AMPL/CPLEX ($c_{ijkt}^t \in [0; 15]$)

$|N|$	$|K|$	Avg CPU time [s]
3	1	0.02
3	3	0.02
4	15	0.09
5	3	0.03
10	3	0.05
15	3	0.04
20	3	0.08
25	3	0.19
50	3	1.42
100	3	2.88

relaxation based heuristic. Due to the aforementioned reasons, we tested only the variants h1, h3, and h6 for our Lagrangean relaxation based procedure (see Sect. 7.2.1). The tests were run on Intel Celeron hardware with 2 GHz and 256 MB RAM. We measure the performance of our heuristical procedure with the average gap of the lower bound LB from the optimum objective function value Opt

$$\text{\%-Gap(LB)} = 100 \cdot \frac{(Opt - LB)}{Opt}$$

and the average gap of the upper bound UB from the optimum objective function value Opt

$$\text{\%-Gap(UB)} = 100 \cdot \frac{(UB - Opt)}{UB} \tag{7.45}$$

and, additionally, with the average number of iterations and the average run-time performance. Table 7.8 presents the results for a different number $|K|$ of products. For results regarding other transportation cost coefficient intervals, compare Table A.1 in the appendix. It is obvious that the lower bounds are extremely better for instances with more products. This effect can be explained by the following fact: The more products are taken into account, the more of these products have an optimum lot sizing plan for the cooperative CLSP that is identical to the uncapacitated version of the lot sizing problem. Therefore, one can assume that capacitated lot sizing problems with fewer products are harder to solve than capacitated lot sizing problems with many products.

Tables 7.9, 7.10, and 7.11 show that our Lagrangean relaxation based procedure can solve larger problems than those of Sambasivan and Yahya (2005) who solved

Table 7.8 CLSP game: average results Lagrangean relaxation based heuristic for a varying number of products

	$|N|$	$|K|$	$[\underline{c}^t_{ijkt}; \overline{c}^t_{ijkt}]$	Avg %-Gap(LB)	Avg %-Gap(UB)	Avg # iterations	Avg CPU time [s]
h1	4	3	[0;15]	4.13%	0.04%	242.00	27.49
	4	15	[0;15]	0.38%	0.06%	282.00	149.90
h3	4	3	[0;15]	4.13%	0.26%	241.47	19.50
	4	15	[0;15]	0.37%	0.11%	254.73	117.97
h6	4	3	[0;15]	4.13%	19.90%	236.00	24.54
	4	15	[0;15]	0.36%	20.04%	231.27	127.23

Table 7.9 CLSP game: average results Lagrangean relaxation based heuristic variant h1 for a varying number of players

$|N|$	$|K|$	$[\underline{c}^t_{ijkt}; \overline{c}^t_{ijkt}]$	Avg %-Gap(LB)	Avg %-Gap(UB)	Avg # iterations	Avg CPU time [s]
3	1	[0;15]	20.30%	0.00%	237.40	13.15
3	3	[0;15]	3.51%	0.04%	221.60	23.45
5	3	[0;15]	5.08%	0.05%	255.80	34.93
10	3	[0;15]	7.70%	0.02%	295.27	65.06
15	3	[0;15]	11.25%	0.04%	307.00	102.78

Table 7.10 CLSP game: average results Lagrangean relaxation based heuristic variant h3 for a varying number of players

$|N|$	$|K|$	$[\underline{c}_{ijkt}^t ; \overline{c}_{ijkt}^t]$	Avg %-Gap(LB)	Avg %-Gap(UB)	Avg # iterations	Avg CPU time [s]
3	1	[0;15]	20.30%	0.00%	240.00	7.20
3	3	[0;15]	3.50%	0.27%	226.00	17.60
5	3	[0;15]	5.08%	0.22%	266.87	24.44
10	3	[0;15]	7.72%	0.16%	289.60	33.63
15	3	[0;15]	11.24%	0.21%	311.13	75.54

Table 7.11 CLSP game: average results Lagrangean relaxation based heuristic variant h6 for a varying number of players

$|N|$	$|K|$	$[\underline{c}_{ijkt}^t ; \overline{c}_{ijkt}^t]$	Avg %-Gap(LB)	Avg %-Gap(UB)	Avg # iterations	Avg CPU time [s]
3	1	[0;15]	20.30%	8.61%	238.00	12.16
3	3	[0;15]	3.47%	17.71%	223.00	22.24
5	3	[0;15]	4.99%	21.40%	257.87	30.55
10	3	[0;15]	7.75%	20.88%	290.20	41.83
15	3	[0;15]	11.26%	20.39%	306.60	82.67

instances with up to four players only. For results regarding other transportation cost coefficient intervals, compare Tables A.2, A.3, and A.4 in the appendix. Furthermore, the results show that the lower bounds get worse with an increasing number of players. On the contrary, the upper bounds seem to be rather stable with a varying number of players. The results allow a comparison of the three tested heuristical variants: Variant h1 provides the best lower and upper bounds but this has its price in the computation time. Compared to the computational time of determining optimal solutions, the Lagrangean relaxation based approaches have a decidedly larger run-time effort (see Table 7.7). Hence, if a state-of-the-art MIP solver (like CPLEX) is available, it is not advisable to apply a Lagrangean relaxation based heuristic to the cooperative CLSP as suggested by Sambasivan and Yahya (2005).

7.4.2 Computational Study: Fix-and-Optimize Heuristic

The tests for the fix-and-optimize procedure were run on an Intel Pentium with 2.8 GHz and 504 MB RAM. If there is no other comment, we used the decomposition variant player based (plb) first, then product based (prb), and at last time based (tb) within each iteration of the fix-and-optimize algorithm. For the time based decomposition we used $\Delta T = 1$ and $\overline{T} = 2$.

The left side of Fig. 7.3 shows the development of the gap (according to (7.45)) over all decompositions and iterations: For $|N| = 25$, $|K| = 3$, $T = 6$, $\Delta T = 1$, and $\overline{T} = 2$, there are 25 player based steps, three product based steps, and five time based steps per iteration; i.e., 33 decomposition steps per iteration. The figure displays average values over 15 instances. The algorithm starts with an average

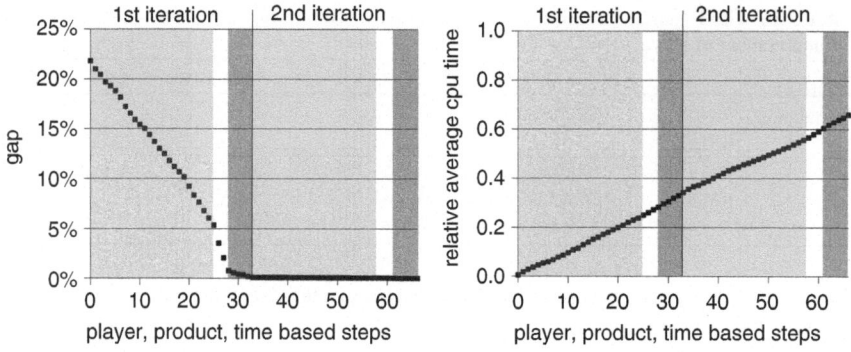

Fig. 7.3 CLSP game: average results for fix-and-optimize (plb + prb + tb with $|N| = 25$ and $c_{ijkt}^t \in [0; 15]$)

Table 7.12 CLSP game: average results fix-and-optimize heuristic for a varying number of players (plb + prb + tb with $c_{ijkt}^t \in [0; 15]$)

$|N|$	$|K|$	Avg %-Gap(UB)	Avg # iterations	Avg CPU time [s]
3	1	0.00%	2.00	0.27
3	3	0.21%	2.47	0.48
4	15	0.02%	3.13	1.94
5	3	0.01%	2.33	0.58
10	3	0.02%	2.40	1.06
15	3	0.04%	2.80	1.92
20	3	0.03%	2.73	2.54
25	3	0.01%	3.13	3.89
50	3	0.01%	3.47	19.31
100	3	0.02%	3.73	284.98

gap of 22% in the first step (solution of the initializing step). The gap decreases during the player based decomposition nearly linear up to an average gap of 5%. It is obvious that the three product based steps bring much more improvement than each of the 25 player based steps before. The procedure reaches a gap smaller than 0.1% with the end of the first iteration. The right diagram of Fig. 7.3 presents the relative CPU time over two iterations (=66 steps). The results show average values over 15 instances with $|N| = 25$ and $c_{ijkt}^t \in [0; 15]$ for the variant plb + prb + tb. The computation time stays nearly the same over all iterations. Coming along with this, Table 7.12 displays the average gap and the average number of iterations for a different number of players. It can be seen that the fix-and-optimize approach is decidedly faster than the Lagrangean relaxation based approach which allows us to solve much larger instances (up to 100 players). For instances up to 15 players, the fix-and-optimize approach yields equally good upper bounds as the Lagrangean relaxation based heuristic variant h1. For very big coalitions the algorithm finds the optimal solution within less than four iterations in average. The number of needed iterations increases slightly with growing coalition size.

Table 7.13 CLSP game: average results fix-and-optimize heuristic for varying transportation cost coefficients (plb + prb + tb, * response time more than 3 h)

$[\underline{c}^t_{ijkt}; \overline{c}^t_{ijkt}]$	$\|N\|$	$\|K\|$	Avg %-Gap(UB)	Avg # iterations	Avg CPU time [s]	$\|N\|$	$\|K\|$	Avg %-Gap(UB)	Avg # iterations	Avg CPU time [s]
[0; 0]	4	15	0.08%	2.87	1.43	25	3	0.06%	2.87	2.66
[0; 5]	4	15	0.04%	3.33	2.09	25	3	0.04%	2.67	2.93
[0; 15]	4	15	0.02%	3.13	1.94	25	3	0.01%	3.13	3.89
[0; 50]	4	15	0.02%	2.93	1.92	25	3	0.01%	3.07	4.56
[5000; 5000]	4	15	0.00%	2.00	1.33	25	3	0.00%*	2.20*	140.07*

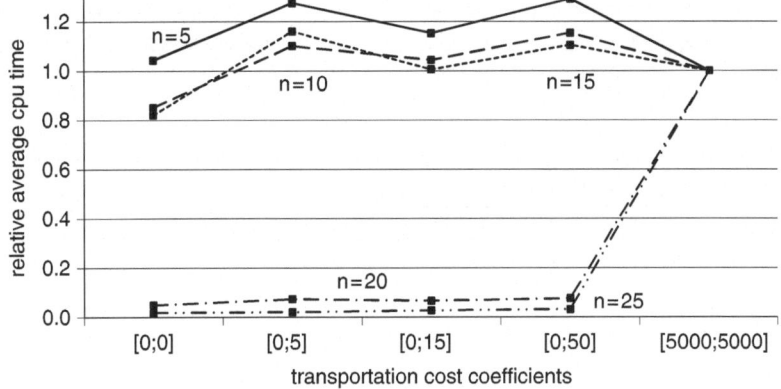

Fig. 7.4 CLSP game: average relative CPU time for varying transportation cost coefficients

One dominating influence for cooperating and producing for another coalition partner is the transportation cost coefficient. Therefore, we analyzed the fix-and-optimize procedure for different transportation cost coefficients. Table 7.13 demonstrates average %-Gap(UB), average number of iterations, and average CPU time for the plb + prb + tb procedure. For results regarding other number of players, compare Table A.5 in the appendix. In case of bigger cooperations ($|N| \geq 20$), the run-time grows decidedly if larger transportation costs are assumed. For a smaller number of players, the run-time is more or less stable.

Figure 7.4 displays the relative computation time. For up to ten players, the computation time does not vary a lot. However, the computation time explodes for bigger cooperations ($|N| \geq 20$) and very high transportation costs ([5000;5000]).

It might have an influence on the solution quality if we change the sequence of the three decomposition procedures. In Fig. 7.5, the development of the gap is displayed for the time, product, player based procedure: Comparing this with Fig. 7.3, that procedure combination seems to reach the optimal solution much faster (with less decomposition steps) than the combination used before. One time based decomposition reduces the gap more than one player based decomposition because the number of optimized variables in the time based decomposition is much higher. Nevertheless, the variant time based first reaches a slightly higher gap after the first

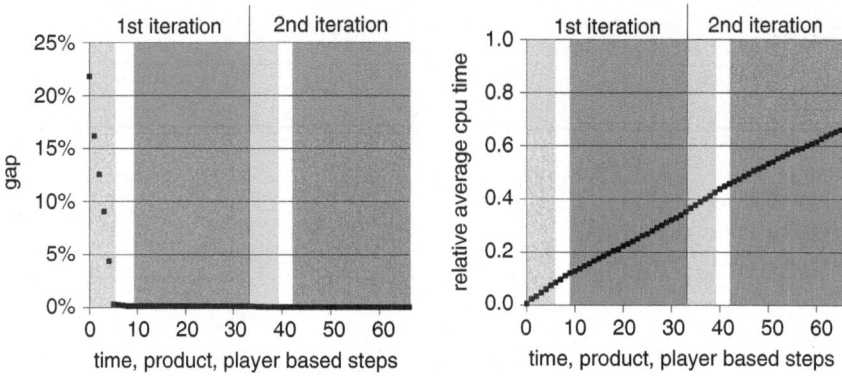

Fig. 7.5 CLSP game: average results fix-and-optimize (tb + prb + plb with $|N| = 25$)

Table 7.14 CLSP game: average results fix-and-optimize heuristic for a varying number of players (tb + prb + plb with $c_{ijkt}^t \in [0; 15]$)

$|N|$	$|K|$	Avg %-Gap(UB)	Avg # iterations	Avg CPU time [s]
3	1	0.00%	2.00	0.27
3	3	0.10%	2.27	0.40
4	15	0.03%	3.27	1.99
5	3	0.01%	2.47	0.59
10	3	0.04%	2.20	0.88
15	3	0.02%	2.87	1.71
20	3	0.05%	2.60	2.36
25	3	0.04%	3.07	4.02
50	3	0.03%	3.07	16.30
100	3	0.02%	4.00	292.41

iteration compared to the variant player based first (0.13% on average) – we will discuss this again later on. Table 7.14 shows results for instances with $c_{ijkt}^t \in [0; 15]$ and different sizes of the cooperation for the decomposition variant time, product, player based. However, we can observe that the order in which the three decomposition principles are applied has only a slight impact on the outcome. Therefore, we tested other combinations of the three decomposition variants. Table 7.15 presents results for instances with $|N| = 50$ and $c_{ijkt}^t \in [0; 15]$. Very low gaps (an average of 0.01% at best) can be achieved with several combinations. Also, the required runtime effort varies a lot. The four best performing variants are prb, tb, tb + prb, and prb + tb.

Furthermore, we investigated whether the sequence of players fixed during the player based decomposition influences the quality of the solution. Helber and Sahling (2010) suggest to classify the products by product specific costs (optimum objective function values of single product subproblems) and the share of the overtime costs (the share depends on the product specific capacity use). Due to the fact that our instances were all feasible by construction (see interval for the values of

Table 7.15 CLSP game: average results fix-and-optimize heuristic depending on the decomposition variant ($|N| = 50$ and $c_{ijkt}^t \in [0; 15]$)

Variant	Avg %-Gap(UB)	Avg # iterations	Avg CPU time [s]
plb	1.13%	7.00	31.60
prb	0.69%	4.00	1.63
tb	0.09%	3.67	2.74
plb + prb	0.47%	3.73	17.86
plb + tb	0.07%	3.87	20.04
prb + plb	0.52%	3.67	18.95
prb + tb	0.01%	3.80	4.50
tb + plb	0.06%	3.73	19.20
tb + prb	0.03%	3.27	3.84
plb + prb + tb	0.01%	3.47	19.30
plb + tb + prb	0.04%	3.13	19.45
prb + plb + tb	0.02%	3.33	18.89
prb + tb + plb	0.01%	3.60	20.48
tb + plb + prb	0.03%	3.33	18.15
tb + prb + plb	0.03%	3.07	16.30

R_{it} at the beginning of this section), we can simplify the formula suggested by Helber and Sahling (2010) a little bit. Hence, we sorted the players according to the individual objective function value based on an LP relaxation:

$$Z_i = \sum_{k \in K} \sum_{t=1}^{T} \left(c_{ikt}^s x_{ikt} + c_{ikt}^h I_{ikt} + c_{ikt}^p q_{ikt} + \sum_{j \in N} c_{ijkt}^t a_{ijkt} \right).$$

The player based decompositions start with the player having the highest and the lowest value Z_i, respectively. Table 7.16 shows average %-Gap(UB), average number of iterations, and average CPU time for the decomposition variant player based (including sequencing), product based, time based. The results reveal that the efficiency of the procedure cannot be improved.

To further reduce the run-time, we stopped the procedure after the first iteration. Obviously, the gap will increase but the procedure needs less time. Table 7.17 shows the results for three different decomposition variants. Recall that the gap for the time based variant decreases faster in the beginning. The results in Table 7.17 show that the gap for the complement variant reaches smaller values after the first iteration, in particular for larger instances, though. The variant prb + tb reaches best results – good bounds in short time.

However, it can be seen that CPLEX is still the better choice even if we compare the run-times of the fix-and-optimize procedure after the first iteration with the time for solving the problem optimally with CPLEX (see Table 7.7). Hence, we can conclude that the fix-and-optimize heuristic proposed by Helber and Sahling (2010) for the (standard) CLSP is not an adequate procedure to tackle the cooperative CLSP – at the presence of today's commercial MIP solvers. Apart from that,

Table 7.16 CLSP game: average results fix-and-optimize heuristic with players in sequence (plb+ prb + tb and $c_{ijkt}^t \in [0; 15]$)

$\|N\|$	$\|K\|$	Starting with $\max_{i \in N} Z_i$			Starting with $\min_{i \in N} Z_i$		
		Avg %-Gap(UB)	Avg # iterations	Avg CPU time [s]	Avg %-Gap(UB)	Avg # iterations	Avg CPU time [s]
3	1	0.00%	2.00	0.31	0.00%	2.00	0.28
3	3	0.07%	2.27	0.44	0.17%	2.40	0.42
4	15	0.02%	3.20	2.04	0.00%	3.33	1.99
5	3	0.01%	2.40	0.56	0.00%	2.27	0.55
10	3	0.03%	2.47	0.96	0.06%	2.20	0.87
15	3	0.03%	2.87	1.89	0.02%	2.93	1.80
20	3	0.03%	2.93	2.75	0.02%	2.73	2.44
25	3	0.03%	3.13	4.14	0.03%	3.07	3.88
50	3	0.03%	3.53	19.63	0.02%	3.53	19.15
100	3	0.02%	3.60	272.67	0.03%	3.73	288.15

Table 7.17 CLSP game: average results fix-and-optimize heuristic after the first iteration ($c_{ijkt}^t \in [0; 15]$)

$\|N\|$	$\|K\|$	plb + prb + tb		tb + prb + plb		prb + tb	
		Avg %-Gap(UB)	Avg CPU time [s]	Avg %-Gap(UB)	Avg CPU time [s]	Avg %-Gap(UB)	Avg CPU time [s]
3	1	0.00%	0.17	0.00%	0.16	0.00%	0.11
3	3	0.33%	0.22	0.21%	0.18	0.22%	0.15
4	15	0.06%	0.64	0.08%	0.60	0.06%	0.50
5	3	0.03%	0.23	0.09%	0.25	0.04%	0.17
10	3	0.11%	0.36	0.08%	0.40	0.08%	0.21
15	3	0.05%	0.71	0.11%	0.59	0.08%	0.25
20	3	0.07%	0.99	0.12%	0.94	0.09%	0.33
25	3	0.07%	1.36	0.13%	1.32	0.10%	0.37
50	3	0.08%	5.88	0.13%	5.48	0.11%	1.23
100	3	0.06%	82.91	0.23%	76.20	0.09%	8.04

Helber and Sahling (2010) do not document a comparison of the fix-and-optimize heuristic against commercial software for the standard CLSP but compare it to other state-of-the-art heuristics and outperform them.

7.4.3 Computational Study: Subcoalition-Perfect Core

We implemented the row generation algorithm for computing an element in the subcoalition-perfect core and run the tests on hardware equipped with an AMD Athlon 64X2 Dual Core Processor 4600+ with 2.41 GHz and 1.96 GB RAM. We only used the instances with $c_{ijkt}^t \in [0; 15]$ and made an important adjustment: The demand was chosen as $d_{ikt} \in [0; d^{max}]$ with $d^{max} = 20$ expect for one player. For player $|N|$, we defined $d_{|N|kt} = 1$ which made the occurrence of a non-monotone

Table 7.18 CLSP game: average results subcoalition-perfect core – average number of iterations

$|N|/|K|$	3/1	3/3	5/3	10/3	15/3
$\pi_i \geq 0$	1.73	2.07	6.87	38.73	195.27
z_i'	2.40	2.80	5.87	22.67	145.53

Table 7.19 CLSP game: average results subcoalition-perfect core – average run-time (CPU-seconds)

$|N|/|K|$	3/1	3/3	5/3	10/3	15/3
$\pi_i \geq 0$	0.02	0.03	0.15	2.84	85.08
z_i'	0.02	0.05	0.13	2.62	109.66

Table 7.20 CLSP game: average results subcoalition-perfect core – average ratio of run-time (CPU-seconds) and number of iterations

$|N|/|K|$	3/1	3/3	5/3	10/3	15/3
$\pi_i \geq 0$	0.01	0.01	0.02	0.07	0.43
z_i'	0.01	0.02	0.02	0.11	0.80

characteristic function very probable (compare the argumentation regarding proving subadditivity for the CLSP game, p. 94).

All our random instances with ten or more players turned out to be non-monotone and most of the instances with smaller number of players are non-monotone as well. Table 7.18 presents the results in terms of the average number of iterations. The average run-time performance measured in CPU-seconds is provided in Table 7.19. All instances were tested for the row generation algorithm with positive π_i-values (as introduced in Sect. 3.5, (3.13)–(3.17)) and for the subproblem formulation (7.37)–(7.42). Table 7.20 displays the time requirement per iteration. It can be observed that computational effort measured in computation time is larger for the new subproblem formulation in comparison to simply using non-negative core allocations. In contrast to that, using the restated subproblem leads to a lower average number of iterations.

7.4.4 Computational Study: Minmax Core

For computing cost allocations in the minmax core, we used the instances with $c_{ijkt}^t \in [0; 15]$ as well and implemented the row generation procedure in AMPL/CPLEX. The tests were run on Intel Pentium hardware with 2.8 GHz and 504 MB RAM.

We observed that the row generation algorithm for the CLSP game is significantly slower than for the ELS game but, however, the number of generated stability constraints is also much lower than the number of constraints that define the core.

Table 7.21 CLSP game: average results minmax core – average number of iterations/average percentage of required constraints (minimum/maximum number of iterations)

$|N|$	$|K|$	$MP(\mathcal{S})$	$MP^I(\mathcal{S})$	$MP^{II}(\mathcal{S})$	$MP^M(\mathcal{S})$
3	1	2.2/70.00%	2.1/67.78%	1.4/56.67%	3.9/97.78%
		(2/3)	(1/3)	(1/3)	(2/4)
3	3	2.4/73.33%	1.3/55.56%	1.1/51.11%	4.0/100.00%
		(2/3)	(1/2)	(1/2)	(4/4)
5	3	4.5/28.22%	2.1/20.44%	1.3/17.56%	6.7/35.78%
		(4/7)	(1/4)	(1/3)	(6/9)
10	3	18.3/2.67%	2.9/1.17%	1.4/1.02%	23.9/3.22%
		(9/40)	(1/7)	(1/4)	(18/30)
15	3	84.4/0.30%	6.8/0.06%	3.3/0.05%	45.5/0.18%
		(31/162)	(1/16)	(1/10)	(19/52)

To compare it with the results for the minmax core, Table 7.21 displays the average number of iterations (over 15 instances), the average percentage of required constraints (see (5.17), p. 74), and the minimum and maximum number of iterations for the classical master problem $MP(\mathcal{S})$, the two fairness variants $MP^I(\mathcal{S})$ and $MP^{II}(\mathcal{S})$, and the minmax core $MP^M(\mathcal{S})$. For all instances, the subproblem variant $\hat{SP}(\pi)$ was used. The efficiency of the approach is proved by the percentage values in Table 7.21. As for the ELS game, the number of constraints actually needed compared to the number of existing constraints measured in percentages leads towards zero with an increasing number of players. In our tests, the run-time effort was within minutes.

Table 7.22 shows the η-values that define the minmax core in our instances. Depending on the $|N|$ and $|K|$ combination, the table provides the minimum, maximum, and average η observed for the 15 instances. For all instances in our test-bed, we could compute cost allocations that are strictly better than the standalone cost for every coalition in the considered game; i.e., $\eta < 1$ for all instances and hence, the core was not empty for the instances in the test-bed. We can conjecture that the core of the cooperative CLSP is not empty in general. Even if this is not true, the minmax core solution concept is able to detect such instances ($\eta > 1$) and suggests a cost assignment anyhow. With an increasing number of players, the minmax core defining η grows (we conjecture that η tends towards one). On the other hand, the results show that all observed η-values are rather close to one which means that there is always at least one coalition in the CLSP game that receives a benefit that is comparatively small in terms of percentages. In other words, when discussing the matter of cooperation for such CLSP settings, it is not possible to promise huge relative benefits to all coalitions. However, with the η-value, it is possible to make some kind of promise on the basis of a minmax core cost allocation which would not be possible on the basis of an arbitrary core cost allocation. In general, a core allocation guarantees cost assignments that are not worse than the standalone situation. Compared to that the promise of at least a small improvement is preferable.

Table 7.22 CLSP game: average results minmax core regarding η – minimum, maximum, and average η

$|N|$	$|K|$	min η	max η	Avg η
3	1	0.9311	0.9898	0.9677
3	3	0.8670	0.9322	0.9075
5	3	0.9230	0.9492	0.9377
10	3	0.9527	0.9753	0.9636
15	3	0.9678	0.9823	0.9775

7.5 Extensions for the CLSP Game

Possible extensions for the presented CLSP game with transshipments are for instance to allow backlogging or include overtime. Such extensions can be included easily. However, the example will show that in case of cooperation new aspects should be taken into account.

We will include backlogging into the model assumptions. This means that demand can be delayed and fulfilled in a later period. We introduce a new cost coefficient c_{ikt}^b and a new decision variable B_{ikt} denoting the backlog quantity of product k in period t. The model for a single decision maker can be formulated as the following *CLSPB_i*:

$$\min \sum_{k \in K} \sum_{t=1}^{T} (c_{ikt}^s x_{ikt} + c_{ikt}^h I_{ikt} + c_{ikt}^b B_{ikt} + c_{ikt}^p q_{ikt}) \qquad (7.46)$$

s.t.

$$I_{ikt} - B_{ikt} = I_{ik,t-1} - B_{ik,t-1} + q_{ikt} - d_{ikt} \qquad k \in K, \, t = 1, \ldots, T \qquad (7.47)$$

$$\sum_{k \in K} q_{ikt} b_{ikt} \leq R_{it} \qquad\qquad t = 1, \ldots, T \qquad (7.48)$$

$$q_{ikt} \leq M_{ikt} x_{ikt} \qquad\qquad k \in K, \, t = 1, \ldots, T \qquad (7.49)$$

$$B_{ikt}, I_{ikt}, q_{ikt} \geq 0 \qquad\qquad k \in K, \, t = 1, \ldots, T \qquad (7.50)$$

$$x_{ikt} \in \{0, 1\} \qquad\qquad k \in K, \, t = 1, \ldots, T. \qquad (7.51)$$

The inventory balance constraints (7.47) have to be extended because in the case of backlogging, demand can be satisfied with production (q_{ikt}), from stock ($I_{ik,t-1}$), or can be backlogged (B_{ikt}) in a following period. The cost parameter for backlogged quantities c_{ikt}^b can be used to control the amount of backlogging (quantity and duration): If c_{ikt}^b is high, then B_{ikt} tends to be smaller. If we set c_{ikT}^b to a big number, then it is assured that there are no backorders at the end of the planning horizon. Obviously, the situation from before without backlogging is a special case with $c_{ikt}^h \ll c_{ikt}^b$.

The cooperative version of the CLSP with backlogging can be formulated while including transshipments:

$$c(S) = \min \sum_{i \in S} \sum_{k \in K} \sum_{t=1}^{T} \left(c_{ikt}^s x_{ikt} + c_{ikt}^h I_{ikt} + c_{ikt}^b B_{ikt} + c_{ikt}^p q_{ikt} + \sum_{j \in S} c_{ijkt}^t a_{ijkt} \right)$$
(7.52)

s.t.

$$I_{ikt} - B_{ikt} = I_{ik,t-1} - B_{ik,t-1} + q_{ikt}$$
$$+ \sum_{j \in S} a_{jikt} - d_{ikt} - \sum_{j \in S} a_{ijkt} \quad i \in S,\, k \in K,\, t = 1,\dots,T \quad (7.53)$$

$$\sum_{k \in K} q_{ikt} b_{ikt} \le R_{it} \qquad\qquad i \in S,\, t = 1,\dots,T \quad (7.54)$$

$$q_{ikt} \le M_{kt} x_{ikt} \qquad\qquad i \in S,\, k \in K,\, t = 1,\dots,T \quad (7.55)$$

$$I_{ik,t-1} + q_{ikt} \ge \sum_{j \in S} a_{ijkt} \qquad i \in S,\, k \in K,\, t = 1,\dots,T \quad (7.56)$$

$$a_{ijkt} \ge 0 \qquad\qquad i,j \in S,\, k \in K,\, t = 1,\dots,T \quad (7.57)$$

$$B_{ikt}, I_{ikt}, q_{ikt} \ge 0 \qquad\qquad i \in S,\, k \in K,\, t = 1,\dots,T \quad (7.58)$$

$$x_{ikt} \in \{0,1\} \qquad\qquad i \in S,\, k \in K,\, t = 1,\dots,T. \quad (7.59)$$

In a cooperative setting, it might happen that it is more cost efficient to fulfill demand of player i for product k while player j backlogs this product. To avoid backlogging for other coalition members, we add restrictions (7.56) which say that transportation quantity of product k in period t from player i to other players $j \in S$ cannot exceed inventory plus production quantity of product k in period t belonging to player i.

Chapter 8
A Multilevel Lot Sizing Game with Restricted Cooperation

In Chap. 5, we started with a very simple form of a purchasing alliance (a dynamic lot sizing problem was the basis for the ELS game). We then extended the scope to cooperative production where scarce capacities occur and transshipments can be used to allocate production in the coalition in a cost optimal way (see Chap. 7). The next step would be to widen the view for multilevel settings that appear in realistic situations in supply networks.

8.1 Cooperative Supply Situations

8.1.1 The Underlying Problem

Consider a supply network composed of several levels (tiers) where each level consists of several players (suppliers, producers, warehouses, customers, etc.). Based on bilateral agreements, the players are connected via transportation channels on which raw materials, intermediate or finished products (depending on the level) can be transported to specific (not necessarily all) players on the following level. Figure 8.1 shows an example network. The circles represent the players that are organized on three levels – imagine players 1 to 4 as suppliers, 5 to 7 as production sites, and 8 and 9 as distributors, for instance. The arcs display the transportation channels.

To keep it simple for the beginning, we assume only one kind of end product and that on every level in the supply network only one (preliminary) product is produced. The end product needs a specific number of preliminary products that can be produced not only by one but several players on the previous level(s) in the network. Comparing Fig. 8.1, the end product produced by player 8 needs one preliminary product that can be delivered by players 5, 6, or 7 (or in combination). Thus, it must be decided for every level which player(s) produce the (preliminary) product.

Continuing the explanation of the previous chapter, we assume that every player faces a capacitated lot sizing problem (CLSP), but with only one product (the

J. Drechsel, *Cooperative Lot Sizing Games in Supply Chains*, Lecture Notes
in Economics and Mathematical Systems 644, DOI 10.1007/978-3-642-13725-9_8,
© Springer-Verlag Berlin Heidelberg 2010

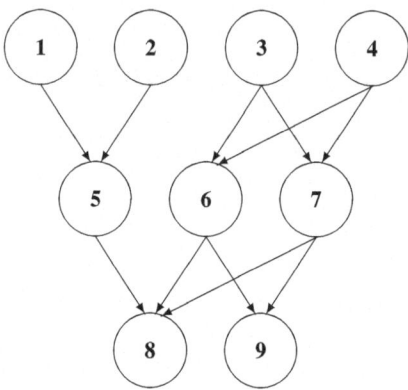

Fig. 8.1 MLCLSP game: example supply network with nine players

following explanations can be easily extended to a multi-product setting). That means, it has to be decided about the period of production, the produced quantity, and the inventory levels under the aim of minimizing total costs (for setup, production and holding) and respecting capacity constraints (see Chap. 7 for more details).

Another aspect from the cooperative CLSP are transshipments: For instance, on the second level in our example network (see Fig. 8.1), transshipments between players 6 and 7 might occur. Both players can deliver preliminary products to player 9. There might be two situations where transshipments are reasonable: Player 6 produces the product, the product needs to be stored for some time, and holding plus transshipment costs for player 7 are smaller than holding costs of player 6. The second situation occurs when player 6 has lower production cost (including transshipment cost) than player 7 but cannot deliver products directly to player 9 (i.e., there is no arc connecting players 6 and 9). In this case, products would be produced by player 6, transshipped to player 7, and the latter delivers the products to player 9. In most of the cases, though, transshipments are included indirectly in our setting, because a preliminary product can be procured by different players and, therefore, they can exchange/pool their capacities as in Chap. 7. We will skip direct transshipments in-between one level for now.

The assumptions can be summarized as follows:

- Players are organized in a multilevel supply network.
- Only one kind of finished product is considered.
- The same (intermediate) product is produced by the players belonging to the same level in the supply network.
- The usage of intermediate products can vary within one level (production coefficient).
- Every player faces a capacitated lot sizing problem.
- Products can only be stored on that level where they were produced.

- Products can only be shipped from one level to another (not within one level).
- Products will be shipped just-in-time and only via given transportation channels to the next level.

We use the following notation for the mathematical formulation: The set N contains all the players $\{1, \ldots, |N|\}$. We introduce a source 0 before the first production level $(N \cup \{0\} = N^0)$. The set F_i contains all players following directly (i.e., on the next downstream level) player i $(i = 0, \ldots, |N|)$. With this general formulation, it is possible to represent situations where levels can be overleaped. As before, T indicates the number of considered periods and c_{it}^s, c_{it}^h, c_{it}^p, and c_{ijt}^t the costs for setup, inventory, production, and transportation. R_{it} still represents the capacity of player i in period t and b_{it} the capacity usage if player i produces its product in period t.

p_{ij} is the production coefficient and indicates how many (intermediate) products produced by player i are needed by player j to produce its product. Let $\bar{p}_{ij} = \frac{1}{p_{ij}}$ for all $j \in F_i$, and $p_{ij} = \bar{p}_{ij} = 0$ for all $j \notin F_i$.

The decision variables are already known from the previous chapters: production quantity of player i in period t (q_{it}), inventory level of player i in period t (I_{it}), setup coefficient for player i in period t (x_{it}) and the transportation quantity from player i to player j in period t (a_{ijt}).

The above described problem can be formalized as follows:

$$\min \sum_{i \in N} \sum_{t=1}^{T} \left(c_{it}^s x_{it} + c_{it}^h I_{it} + c_{it}^p q_{it} + \sum_{j \in N} a_{ijt} c_{ijt}^t \right) \qquad (8.1)$$

s.t.

$$I_{it} = I_{i,t-1} + q_{it} - \sum_{j \in F_i} a_{ijt} - d_{it} \qquad i \in N, t = 1, \ldots, T \quad (8.2)$$

$$q_{it} = \sum_{h \in N^0} a_{hit} \bar{p}_{hi} \qquad i \in N, t = 1, \ldots, T \quad (8.3)$$

$$q_{it} b_{it} \leq R_{it} \qquad i \in N, t = 1, \ldots, T \quad (8.4)$$

$$q_{it} \leq M x_{it} \qquad i \in N, t = 1, \ldots, T \quad (8.5)$$

$$I_{it}, q_{it} \geq 0 \qquad i \in N, t = 1, \ldots, T \quad (8.6)$$

$$x_{it} \in \{0, 1\} \qquad i \in N, t = 1, \ldots, T \quad (8.7)$$

$$a_{ijt} \geq 0 \qquad i \in N^0, j \in N, t = 1, \ldots, T. \quad (8.8)$$

The objective function (8.1) minimizes total costs for the whole network (costs for setup, inventory, production, and transportation for all players). The inventory balances are given by constraints (8.2) that include production quantity of the product, transported items to the following level in the network, and external demand for the (intermediate) product. Equations (8.3) assure that corresponding to p_{hi} enough preliminary products from other players $(h = 0, \ldots, |N|)$ are transported to player

i for producing q_{it}. Due to the fact that on one level only one (preliminary) product is produced, the sum over transportation quantities from all direct predecessors in the network multiplied with the corresponding production coefficient must equal the production quantity of player i. Restrictions (8.4) and (8.5) are the well known capacity constraints and the setup constraints. M is a sufficiently big number and needs to be at least as high as the external demand of all players plus the secondary demand resulting from the external demand through the whole network. At last, (8.6)–(8.8) are the domains for the decision variables.

Due to constraints (8.3), we can substitute q_{it} by $\sum_{h \in N^0} (a_{hit} \bar{p}_{hi})$ and rewrite the model:

$$\min \sum_{i \in N} \sum_{t=1}^{T} \left(c_{it}^s x_{it} + c_{it}^h I_{it} + \sum_{h \in N^0} a_{hit} \bar{p}_{hi} c_{it}^p + \sum_{j \in N} a_{ijt} c_{ijt}^t \right) \qquad (8.9)$$

s.t.

$$I_{it} = I_{i,t-1} + \sum_{h \in N^0} a_{hit} \bar{p}_{hi} \sum_{j \in F_i} a_{ijt} - d_{it} \qquad i \in N, \, t = 1, \ldots, T \quad (8.10)$$

$$b_{it} \sum_{h \in N^0} a_{hit} \bar{p}_{hi} \le R_{it} \qquad i \in N, \, t = 1, \ldots, T \quad (8.11)$$

$$\sum_{h \in N^0} a_{hit} \bar{p}_{hi} \le M \, x_{it} \qquad i \in N, \, t = 1, \ldots, T \quad (8.12)$$

$$I_{it} \ge 0 \qquad i \in N, \, t = 1, \ldots, T \quad (8.13)$$

$$x_{it} \in \{0, 1\} \qquad i \in N, \, t = 1, \ldots, T \quad (8.14)$$

$$a_{ijt} \ge 0 \qquad i \in N^0, \, j \in N, \, t = 1, \ldots, T. \quad (8.15)$$

Instead of restrictions (8.12), it is possible to use a disaggregated formulation:

$$a_{hit} \le M \, x_{it} \qquad h \in N^0, \, i \in N, \, t = 1, \ldots, T.$$

It simply says that if any preliminary product is shipped from player h to player i, they are used to make new products and, therefore, a setup must take place (a production step takes place whenever a product reaches a new player).

As in the previous chapters concerning the ELS game and the CLSP game, we will speak of a cooperation if several players work together to produce something. The main difference now is that we need a minimal set of players to produce the demanded product. We define a set of $P_i \subset N$ for every player i containing its direct upstream players:

$$P_i = \{j \mid p_{ji} > 0, \, j \in N\}.$$

If one player i (or a subset $S' \subset N$ of players) in the coalition S has external demand ($\sum_{t=1}^{T} d_{it} > 0$, $i \in S'$), we need at least one player out of its predecessors in P_i ($i \in S'$) (depending on the capacity) to provide the upstream pre-products:

A directed path L_g^e is a sequence of connected players starting with the source 0 and ending with player e:

$$0 = i_1 - \ldots - i_l = e \quad \text{for all } e \in N \text{ and } j = 1 \ldots, l-1 \quad \text{where } p_{i_j i_{j+1}} = 1.$$

Define a set L^e that contains all paths g leading to player e and G denotes the number of such paths:

$$L^e = \{L_g^e \mid g = 1, \ldots, G\} \quad \text{for all } e \in N.$$

Then $\tilde{F}$ is a union of paths:

$$\tilde{F} \subseteq \bigcup_{e \in N} L^e$$

and a set $S \subseteq N$ is called a feasible coalition if

$$\forall i \in S : \left(\sum_{t=1}^{T} d_{it} > 0 \Rightarrow \exists f \in \tilde{F} \; \forall j \in f : j \in S \right).$$

The grand coalition N and the empty coalition $\emptyset$ are feasible coalitions as well. Assume for our example in Fig. 8.1 that player 9 has primary demand. Hence, a coalition containing players 9 and 7 is not enough to form a coalition because player 7 cannot produce its product since there is no one in the coalition who can provide the preliminary product for player 7. A coalition containing only one single player or containing players from just one level is only feasible if the players in such a coalition have no external demand or no upstream players (no pre-products). $S = \{6\}$ would only be feasible if this player has no external demand and, thus, does not need pre-products. Otherwise it could only work in coalitions with player 3 and/or 4. Actually, it is not single players cooperating but paths or subnetworks.

For building up a cooperative multilevel capacitated lot sizing game (MLCLSP game), there must be an incentive for cooperation among the players. The motivation comes from

- the necessity that preliminary products are needed to produce end products and
- sharing/pooling production capacities among the players on one level.

Hence, there is an important difference between the game described above and the cooperative situations in the previous chapters: The set of feasible subcoalitions is restricted – the characteristic function is not defined on the collection of all subsets of players anymore. Therefore, we will review cooperative games with restricted cooperation in the next section.

8.1.2 Games with Restricted Cooperation

We call $\mathcal{F}$ the family of feasible coalitions $S \subseteq N$. Accordingly, $\Gamma = (\mathcal{F}, v)$ is called a cooperative game with restricted cooperation, where v is the characteristic function.

The literature covers different models for restricted cooperation: Myerson (1977) uses *graph*-theoretic ideas: The cooperating players are connected by links (one link is connecting two players), representing bilateral cooperative agreements. Hence, the graph containing the players and the links shows us the cooperation structure. Myerson (1977) proposes an allocation rule that is mostly based on the links the players belong to and not the players themselves. Hence, a player should get a higher payoff if its position is more essential to coordinate the other players (e.g., this player is connected to all other players and the other players have no direct connection among themselves). Further contributions to the topic of graph-restricted games are made by Owen (1986), Borm et al. (1992), and Hamiache (1999). This approach is not applicable for our MLCLSP game because not every subcoalition containing connected players is feasible (e.g., if pre-products are missing).

Cooperative games with *precedence constraints* on the players are studied by Faigle and Kern (1992). With these constraints, the set of players is (partially) ordered by some precedence relation. A feasible coalition is characterized by the following property: If player i is member of a coalition S, then all players preceding i must be members of the same coalition as well. Among investigating the Shapley value for cooperative games under precedence constraints, Faigle and Kern (1992) show that multichoice games (see Sect. 2.2.3) can be transformed into games under precedence constraints (the individual player's actions form the relevant set of players). Several more articles (e.g., Gilles et al. 1992; Derks and Gilles 1995; van den Brink 1997; Gilles and Owen 1999) deal with games with *permission structures* – a related concept to precedence constraints. Here, a hierarchical organizational structure is used where each player has at least one chain of superiors that has to give permission to the action of this player. A collection of all permission structures is defined by $\mathcal{S}(N)$. $S \in \mathcal{S}(N)$ describes a hierarchical structure on N, where $j \in S(i)$ is interpreted as that player i dominates player j. Hence, player i has veto power over all actions undertaken by its subordinates j. Our MLCLSP game could be interpreted as a hierarchical structure. Take the example of Fig. 8.1: Player 9 needs permissions by players 6 or 7 and 3 or 4 to take part in the coalition and produce something due to the need of pre-products. However, the difference is that player 9 needs permission from player 6 or 7 and not both.

Derks and Gilles (1995) already make a transition from such hierarchical structures to so called *lattices*. Furthermore, Grabisch and Lange (2007) and Grabisch and Xie (2007) discuss games on lattices which are based on hierarchical structures and precedence constraints. The characteristic function is considered as a real-valued function defined over a (often distributive) lattice. This approach can, e.g., be used for a problem like sharing the total benefit of a company with N employees who are structured in a hierarchy and form teams among themselves. Again, such teams cannot be predefined in the case of the MLCLSP game because it is unknown which players from the upstream levels belong to the team (i.e., are producing pre-products in the subcoalition).

Bilbao et al. (1999) develop a more general coalition structure using the theory of *convex geometry*. In this approach, the collection of feasible coalitions is characterized by the following properties:

- The empty set and the grand coalition are feasible coalitions: $\emptyset \in \mathcal{F}$ and $N \in \mathcal{F}$.
- If two coalitions are feasible, then their intersection is it as well: $A \in \mathcal{F}$ and $B \in \mathcal{F}$ implies that $A \cap B \in \mathcal{F}$.
- A feasible subcoalition (not the grand coalition) is still feasible when it cooperates with some external player.

For further details and enhancements concerning convex geometries, see Bilbao (2000), Okamoto (2003), and Bilbao et al. (2006). Bilbao (2000) examines further models of games with restricted cooperation as well. The mentioned papers transfer well known theorems and properties from classical TU games to their concept of games on convex geometries. Particularly, the second property does not hold for our MLCLSP game. For instance, the intersection of the subcoalitions $S_1 = \{3, 6, 9\}$ and $S_2 = \{4, 6, 9\}$ is not feasible due to missing pre-products for player 6 (see Fig. 8.1).

The *antimatroid* is a further more general structure introduced by Algaba et al. (2004): An antimatroid $\mathcal{A}$ on N is a family of subsets of 2^N, satisfying the following assumptions:

- The empty set is a feasible coalition: $\emptyset \in \mathcal{F}$.
- The union of two feasible coalitions is also feasible: If $E, F \in \mathcal{A}$, then $E \cup F \in \mathcal{A}$. (*closed under union*)
- There is one player in a nonempty feasible coalition, so that if he leaves the coalition, it is still feasible: If $E \in \mathcal{A}$, $E \neq \emptyset$, then there exists $i \in E$ such that $E \setminus \{i\} \in \mathcal{A}$. (*accessibility*)
- The grand coalition is a feasible coalition: For every $i \in N$ there exists an $E \in \mathcal{A}$ such that $i \in E$. This implies that $N \in \mathcal{A}$. (*normality*)

Some references distinguish between antimatroids (defined by the first three properties) and normal antimatroids (defined by all four properties) (e.g., Bilbao 2003).

To generalize the concepts of antimatroid structure and graph-restricted games, Bilbao (2003) develops another combinatorial structure called *augmenting systems*. An augmenting system is denoted by the following properties:

- The empty set is a feasible coalition: $\emptyset \in \mathcal{F}$.
- The union of two feasible non-disjoint coalitions is also feasible: If $E, F \in \mathcal{F}$ with $E \cap F \neq \emptyset$, then $E \cup F \in \mathcal{F}$.
- For $E, F \in \mathcal{F}$ with $E \subset F$, there exists $i \in F \setminus E$ such that $E \cup i \in \mathcal{F}$. (*augmentation property*)

For recent advances concerning augmenting systems see also Bilbao and Ordóñez (2009).

The second property for antimatroids and augmenting systems does not need to be true for the MLCLSP game. For instance, the union of the subcoalitions

$S_1 = \{3, 6, 8\}$ and $S_2 = \{3, 6, 9\}$ might not be feasible due to lack of the suppliers' capacities (see Fig. 8.1) providing pre-products for players 8 and 9.

In *partitioning* games, the restricted coalitions are defined over a partition system. A partition system is a set system where

- $\emptyset \in \mathcal{F}$ and $\{i\} \in \mathcal{F}$ for every $i \in N$ and
- if and only if $S_1, S_2 \in \mathcal{F}$ and $S_1 \cap S_2 \neq \emptyset$ imply $S_1 \cup S_2 \in \mathcal{F}$.

Partitioning games are introduced by Kaneko and Wooders (1982) and further studied by Quindt (1991), Bilbao (2000), and Solymosi (2008). Due to the construction regarding single player coalitions, partitioning games are not suitable for our MLCLSP game either.

For the sake of completeness, there exist the concepts of games on *regular set systems* (see Xie 2006) and *fuzzy games* (see Butnariu and Klement 1993). These models are not that much discussed in the literature and not applicable for our problem.

A possibility to avoid a new model for restricted cooperation is to define an extension of the game which then is a classical TU game. Derks and Peters (1993), Faigle and Kern (2000), and Faigle and Peis (2006) present further details about *game extensions*.

Apart from concentrating on different approaches to present restricted cooperation, there are references focusing on special properties of the core of games with restricted coalitions. For instance, Pulido and Sánchez-Soriano (2006) characterize the core for restricted cooperation based on the core characterization developed by Peleg (1999). Derks and Reijnierse (1998) analyze non-degeneracy, balancedness, exactness, etc.

Algaba et al. (2001) present a unified approach while introducing feasible coalition systems, partition systems, intersecting systems, and partition convex geometries and their relationship. Faigle and Peis (2008) develop a general framework subsuming all the before mentioned models for games with restricted cooperation.

Some of the before mentioned papers analyze allocation concepts as well. Faigle and Peis (2008) show how the notion of the *core* introduced in Sect. 2.3.4 can easily be extended to the case of restricted cooperation in profit games. Faigle (1989) points out that the core is also meaningful for situations with restricted cooperation and that concerning the existence of the core the model remains essentially the same if $N \in \mathcal{F}$ and if $v(N)$ is the value for the grand coalition. See Faigle (1989) for proofs about balancedness, corresponding core theorems, and convex games. According to Faigle and Peis (2008), we define the core of a profit game with restricted cooperation as allocation vectors that do not allocate more than v^* in total and each feasible coalition S receives at least its value $v(S)$:

$$C(v) = \left\{ \pi \in \mathbb{R}^{|N|} \,\middle|\, \sum_{i \in N} \pi_i = v^*, \ \sum_{i \in S} \pi_i \geq v(S) \quad \text{for all } S \in \mathcal{F} \right\}.$$

The core has the advantage that an extension to restricted cooperation is much easier than for other allocation methods; e.g., the Shapley value where some game extensions are necessary (see Faigle and Peis 2008). For the formulation of the core, it is no problem if any subcoalition is missing as long as $\mathcal{F} \neq \emptyset$ and the grand coalition is feasible. If $N \notin \mathcal{F}$, we can define the sum of the profit for pairwise disjoint coalitions to generate the total profit (Faigle and Peis 2008):

$$v^* = \max \left\{ \sum_l v(S_l) \,\middle|\, S_l \in \mathcal{F}, \ S_l \cap S_k = \emptyset \quad \text{if } l \neq k \right\}. \tag{8.16}$$

Assume a cooperative profit game with five players $N = \{1, 2, 3, 4, 5\}$ and feasible coalitions $\mathcal{F} = \{\{1, 2\}, \{2, 3\}, \{4, 5\}, \{3, 4, 5\}\}$. According to (8.16), we have

$$\begin{aligned} v^* = \max\{ & v(\{1, 2\}) + v(\{3, 4, 5\}), \\ & v(\{1, 2\}) + v(\{4, 5\}), \\ & v(\{2, 3\}) + v(\{4, 5\})\}. \end{aligned}$$

The total profit for the grand coalition is defined by the highest value for pairwise disjoint sets of feasible subcoalitions. Note, that the added subcoalitions do not need to contain all players but every player at most once.

Most of the literature covering restricted cooperation discusses cooperative profit games. A classical cooperative profit game can be transformed into a cooperative cost game so that both have the same core and vice versa. Consider a profit game (N, v) and a cost game (N, c). The core for both games is described by the following constraints:

cost game:

$$\sum_{i \in N} \pi_i' = c(N)$$

$$\sum_{i \in S} \pi_i' \leq c(S) \quad S \subset N$$

profit game:

$$\sum_{i \in N} \pi_i = v(N)$$

$$\sum_{i \in S} \pi_i \geq v(S) \quad S \subset N$$

Assume $c(S) = -v(S)$ for all $S \subseteq N$.

$$\rightarrow \quad \sum_{i \in N} \pi_i' = -v(N)$$

$$\sum_{i \in S} \pi_i' \leq -v(S) \quad S \subset N$$

$$\rightarrow \quad -\sum_{i \in N} \pi_i' = v(N)$$

$$-\sum_{i \in S} \pi_i' \geq v(S) \quad S \subset N$$

Set $\pi_i = -\pi_i'$ for all $i \in N$.

However, Faigle and Peis (2008) state that this equivalence is no longer true for a cooperative game with restricted cooperation. The core for a cooperative cost game with restricted cooperation can be formulated as follows (see Faigle and Peis 2008):

$$C(c) = \left\{ \pi \in \mathbb{R}^{|N|} \;\middle|\; \sum_{i \in N} \pi_i = c^*, \; \sum_{i \in S} \pi_i \leq c(S) \text{ for all } S \in \mathcal{F} \right\}.$$

Like before, it might be possible that the grand coalition is not feasible. The total costs can be calculated with this equation:

$$c^* = \min \left\{ \sum_l c(S_l) \;\middle|\; N \subseteq \bigcup_l S_l, \; S_l \in \mathcal{F} \right\}.$$

Total costs result from the smallest accumulated costs of non-disjoint subsets containing all players of N. This formulation assures that the costs induced by every player were considered for the grand coalition. This is in contrast to the total profit (8.16), where no player should bring in individual profit more than once. For our numerical example (see p. 127), total costs can be computed with

$$\begin{aligned}
c^* = \min\{ &c(\{1,2\}) + c(\{2,3\}) + c(\{4,5\}) + c(\{3,4,5\}), \\
&c(\{1,2\}) + c(\{3,4,5\}), \\
&c(\{1,2\}) + c(\{2,3\}) + c(\{4,5\}), \\
&c(\{1,2\}) + c(\{2,3\}) + c(\{3,4,5\}), \\
&c(\{1,2\}) + c(\{4,5\}) + c(\{3,4,5\})\}.
\end{aligned}$$

The *unboundedness* of the allocation values is another notable issue when dealing with restricted cooperation: In classical TU profit games the individual rationality conditions suffice to ensure that the allocation values are bounded. If single coalitions are not feasible, there is no lower bound for the corresponding allocation which means that the core is unbounded (Grabisch and Xie 2008). The same holds for cost games with restricted cooperation how the following small example shows. Assume a cooperative game with three players, three feasible coalitions $\mathcal{F} = \{\{1,3\}, \{2,3\}, \{1,2,3\}\}$, and $c(\{1,3\}) = 15$, $c(\{2,3\}) = 15$, $c(N) = 20$. Hence, the core is described by these three constraints:

$$\pi_1 + \pi_3 \leq 15 \tag{8.17}$$
$$\pi_2 + \pi_3 \leq 15 \tag{8.18}$$
$$\pi_1 + \pi_2 + \pi_3 = 20. \tag{8.19}$$

Equation (8.17) in combination with (8.19) yields to $\pi_2 \geq 5$ and (8.18) with (8.19) leads to $\pi_1 \geq 5$ which results in $\pi_3 \leq 10$. Thus, π_1 and π_2 do not have an upper bound and π_3 misses a lower bound. Due to that the core is unbounded. Grabisch and Xie (2008) develop an extension of the core for their games on lattices so that the core is a closed convex polytop (by imposing further normalization constraints). Note that this concept does not work for classical cooperative games where all sub-coalitions are feasible (see Grabisch and Xie 2008). The before mentioned extension of the game is another possibility to make the core bounded. The result is a classical TU game and we can determine the classical core of the extended game.

8.1.3 Properties of the MLCLSP Game

We will now analyze some properties of the MLCLSP game. Most of the reasoning is nearly similar to the CLSP game (see Sect. 7.1.3). The definitions of monotonicity, subadditivity, and concavity are derived from classical cooperative TU games and are now applied on feasible subcoalitions $S \in \mathcal{F}$ and not all $S \subseteq N$.

The MLCLSP game is *monotone* if (2.1) holds for S_1, $S_2 \in \mathcal{F}$ (see p. 11). Compare the example in Fig. 8.1: A feasible subcoalition S contains players 1, 5, and 8. Player 2 has no external demand over the time horizon but free capacity. If player 2 joins the subcoalition S, total costs would not increase because the optimal solution for the coalition S would also be feasible for the coalition $S \cup \{i\}$. Such a solution would be optimal for the coalition $S \cup \{i\}$ if player 1 could produce the intermediate product for lower costs than player 2 or the two together. Otherwise, total costs would decrease which contradicts the monotonicity property. Hence, the MLCLSP game is not monotone in general.

To test whether the players have an incentive to cooperate, *subadditivity* should be verified. The MLCLSP game is subadditive if (2.2) holds for $S_1, S_2 \in \mathcal{F}$ (see p. 12). The supply network game is subadditive because adding the optimal solutions for two feasible subcoalitions S_1 and S_2 provides a feasible solution for the coalition $S_1 \cup S_2$. Therefore, the union of the two feasible coalitions would never have a higher objective function value than the two subcoalitions when working alone. Hence, the MLCLSP game is subadditive and the players have an incentive to cooperate.

Concavity is an important property because if a cooperative cost game is concave, core cost allocations can be computed with simple standard methods (e.g., while using marginal vectors). The MLCLSP game is concave if (2.3) holds for S_1, $S_2 \in \mathcal{F}$. A small counterexample in Sect. 8.2.2 shows that it is not concave in general.

8.2 Computing Core Cost Allocations for the MLCLSP Game

After learning about the MLCLSP game, its specific characteristic of restricted cooperation, the literature contribution to the topic of restricted cooperation, and the properties of such games, let us now analyze in detail how the cooperative MLCLSP game should be treated to compute fair and stable cost allocations.

The literature survey in Sect. 8.1.2 indicates that most of the references regarding restricted cooperation are dealing with convex and monotone games for proofing properties, theorems, and sometimes efficient algorithms. However, to the best of our knowledge, there is no literature dealing with such complex non-monotone and non-concave games including restricted cooperation.

8.2.1 The Row Generation Procedure

Continuing the procedure introduced in Sect. 3.2, we now adjust the proposed row generation algorithm to the MLCLSP game to compute core cost allocations. Recall that the master problem is given by (3.1)–(3.5). As the grand coalition is always feasible, the total costs $c(N)$ can be computed while solving the before presented cooperative MLCLSP in advance (with standard software like CPLEX). Due to the presence of restricted cooperation as described in Sect. 8.1.1, the starting set $S = \{1, \ldots, |N|\}$ for the rationality restrictions as used for the ELS game and the CLSP game cannot be used anymore (because a single player is not necessarily a feasible coalition).

Before formulating the subproblem, one question arises: Who should bear the costs for the production of pre-products – the supplier or the player who finishes the end product? There might be players in the grand coalition that do not have external demand – they are serving only as suppliers of pre-products to other players. Should we assign costs of zero to them?

One might argue that players without external demand would not have an incentive to take part in the cooperation but due to $\pi_i \in \mathbb{R}$, they could get negative cost shares which can be interpreted as profit. Moreover, the row generation procedure computes $c(S)$-values based on the MLCLSP. This means, we do not take into account that pre-products for downstream players that are not contained in this coalition S might be, however, produced in the grand coalition. This is adequate to set $c(S) = 0$ if the coalition S has no external demand. Analog, profits would be shared the same way.

In the following, we assume that the player who induces the costs by means of external demand should pay them. Thus, a player i without external demand ($d_{it} = 0$ for all $t = 1, \ldots, T$) can form a single coalition with $c(\{i\}) = 0$ which leads to the stability constraint $\pi_i \leq 0$. With this knowledge, we add such stability constraints to the master problem before starting the algorithm. Fixing those π_i-values to zero would limit the solution space because of cutting solutions with negative π_i-values. Thus, the pre-processing includes all single coalitions of players who have no external demand in $\mathcal{S}$:

$$\mathcal{S} = \left\{ \{i\} \,\middle|\, \sum_{t=1}^{T} d_{it} = 0, \ i \in N \right\}. \tag{8.20}$$

Results from the master problem provide the starting values for the first iteration of the *subproblem* to generate new rationality constraints for the master problem: $\hat{SP}(\pi)$:

$$\max -o + \sum_{i \in N} \pi_i z_i \tag{8.21}$$

s.t.

$$o = \sum_{i \in N} \sum_{t=1}^{T} \left(c_{it}^s x_{it} + c_{it}^h I_{it} + \sum_{h \in N^0} a_{hit} \bar{p}_{hi} c_{it}^p + \sum_{j \in N} a_{ijt} c_{ijt}^t \right) \tag{8.22}$$

$$I_{it} = I_{i,t-1} + \sum_{h \in N^0} a_{hit} \bar{p}_{hi} - \sum_{j \in F_i} a_{ijt} - d_{it} z_i \qquad i \in N, t = 1, \ldots, T \tag{8.23}$$

$$b_{it} \sum_{h \in N^0} a_{hit} \bar{p}_{hi} \leq R_{it} \qquad i \in N, t = 1, \ldots, T \tag{8.24}$$

$$a_{hit} \leq M_t x_{it} \qquad h \in N^0, i \in N, t = 1, \ldots, T \tag{8.25}$$

$$\sum_{t=1}^{T} \sum_{j \in N} a_{ijt} \leq z_i \sum_{t=1}^{T} M_t \qquad i \in N \tag{8.26}$$

$$I_{it} \geq 0 \qquad i \in N, t = 1, \ldots, T \tag{8.27}$$

$$x_{it} \in \{0, 1\} \qquad i \in N, t = 1, \ldots, T \tag{8.28}$$

$$a_{ijt} \geq 0 \qquad i \in N^0, j \in N, t = 1, \ldots, T \tag{8.29}$$

$$z_i \in \{0, 1\} \qquad i \in N. \tag{8.30}$$

The objective function is already known from the CLSP game (see Sect. 7.3). The target is to maximize the difference between $c(S)$ and the cost shares for this coalition S ($\sum_{i \in S} \pi_i$) to find a coalition violating the rationality restriction most. Constraint (8.22) defines the cost value $c(S)$ ($= o$). The restrictions (8.23)–(8.25) and the domains (8.27)–(8.29) are taken from the MLCLSP game (see Sect. 8.1.1). Decision variables z_i indicate whether a player i belongs to the found coalition ($z_i = 1$) or not ($z_i = 0$). Constraints (8.26) assure that only feasible coalitions are found: If there is a transportation quantity in any period from player i to any player $j \in N$ (i.e., player j produces something), then player i needs to be in the coalition. This implies that if there is no transportation quantity in any period from player i to any player $j \in N$, this player could belong to the coalition anyhow.

Solving the problem for the grand coalition (as formulated in Sect. 8.1.1) can lead to an optimal solution where one or several players do not participate in the cooperative production process (i.e., the corresponding decision variables take values of zero). Due to the non-monotonicity of the characteristic function, these players cannot be excluded from the player set N to simplify the problem since they could participate in a subcoalition. The same holds for players who are not involved in

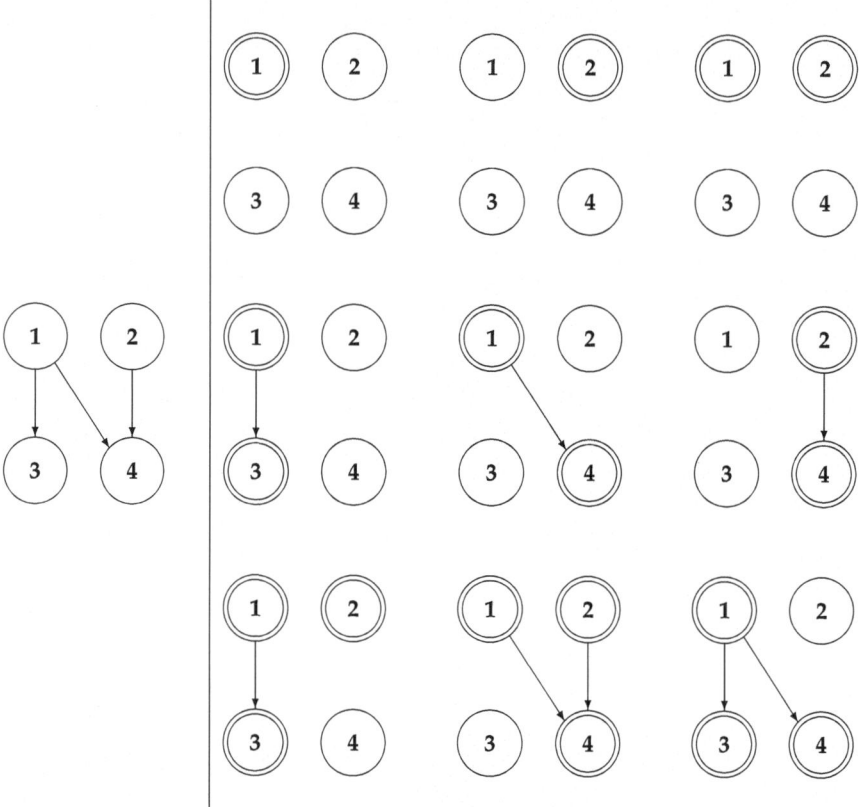

Fig. 8.2 MLCLSP game: example supply network with four players and nine feasible subcoalitions S beside $\emptyset$ and the grand coalition N (players belonging to the subcoalition are marked with *doubled circles*)

the production process of a subcoalition (due to too many costs or the absence of transportation channels).

Figure 8.2 shows an example network with two levels and four players. Players 1 and 3, 1 and 4, as well as 2 and 4 are connected via transportation channels. Note that there is no connection between player 2 and 3. A coalition of the players 1, 2, and 3 is feasible even though player 2 would not be involved in production as long as he has no external demand (assume for the moment that $\sum_{t=1}^{T} d_{2t} = 0$). Such a coalition S can be reduced to a smaller coalition $S \backslash \{i\}$, for the example, $S = \{1, 2, 3\}$ can be reduced to $S \backslash \{2\} = \{1, 3\}$. The value for the characteristic function is $c(S) = c(S \backslash \{i\})$, for the example, $c(\{1, 2, 3\}) = c(\{1, 3\})$. Hence, the subproblem might generate constraints like, for instance,

$$\pi_1 + \pi_2 + \pi_3 \le c(\{1, 3\}) \text{ and}$$
$$\pi_1 + \pi_3 \le c(\{1, 3\}).$$

Table 8.1 MLCLSP game: parameters for the numerical example

d_{it}	$t=1$	2	3	c_{it}^s	1	2	3	c_{it}^h	1	2	3	c_{ijt}^t	$j=1$	2	3	4
$i=1$	5	0	0	1	100	100	100	1	7	7	7	$i=1$	5	5	5	5
2	0	0	0	2	50	50	50	2	5	5	5	2	10	10	10	10
3	10	15	10	3	50	50	50	3	5	5	5	3	10	10	10	10
4	10	15	15	4	100	100	100	4	7	7	7	4	10	10	10	10

The complete row generation algorithm proceeds nearly similar compared to the ELS game or the CLSP game: $\hat{C}ore(MP, \hat{SP})$:

1. Define a small initial set S as proposed in (8.20). Compute the total costs $c(N)$ for the coalition N.
2. Solve the linear program $MP(S)$ (see (3.1)–(3.5)) optimally.
3. If $w > 0$, stop the algorithm because the instance has an empty core.
4. Otherwise, solve $\hat{SP}(\pi)$ (see (8.21)–(8.30)) optimally.
5. If $\hat{SP}(\pi)$ has a non-positive optimum objective function value, then stop the algorithm because the found allocation is in the core.
6. Otherwise, set $c(S') = o$, update $S = S \cup \{S'\}$ and go to Step 2.

The following section presents the whole algorithm for a numerical example.

8.2.2 A Numerical Example

Take again the supply network displayed on the left of Fig. 8.2. The instance has three periods. Table 8.1 shows the parameters for the multilevel production process. The transportation cost coefficients are constant over the time horizon. Furthermore, it is considered for all periods and players $c_{it}^p = 0$, $R_{it} = 30$ and $b_{it} = 2$.

On the right side of Fig. 8.2, the nine feasible subcoalitions are presented. Only two single coalitions are feasible ($\{1\}$ and $\{2\}$) because only those players do not need pre-products. The two-player-coalitions $\{2, 3\}$ and $\{3, 4\}$ are not feasible because in both subcoalitions there is no possibility to produce the pre-products needed by player 3 (and in the second subcoalition for player 4). For the same reason, the three-player-coalition $\{2, 3, 4\}$ is not feasible. The grand coalition is feasible. Note that the coalition $\{1, 3, 4\}$) might not be feasible because production capacities of player 1 could not suffice to produce the pre-products required by players 3 and 4. The capacity restrictions in the subproblem will handle such cases.

Before using the presented row generation algorithm to compute a core cost allocation for this instance, we solve the problem for the grand coalition. Afterwards, we add one stability constraint $\pi_2 \leq 0$ due to $\sum_{t=1}^{T} d_{2t} = 0$.

The row generation procedure is implemented in AMPL and solved with the solver CPLEX 10.0.0. The first master problem provides the simplest feasible solution: $\pi_1 = 1{,}450$ and $\pi_2 = \pi_3 = \pi_4 = 0$. Solving the subproblem, we get

$S = \{1, 2\}$ as subcoalition that violates the core constraints most. Only the primary demand of player 1 needs to be produced in this coalition which induces costs of 100. This generates the next stability constraint for the master problem: $\pi_1 + \pi_2 - v \leq 100$. After two more iterations, the following master problem reveals the core cost allocation $\pi_1 = -25$, $\pi_2 = 0$, $\pi_3 = 650$ and $\pi_4 = 825$:

$$\min w$$

s.t.

$$\pi_1 + \pi_2 + \pi_3 + \pi_4 = 1,450 \tag{8.31}$$
$$\pi_2 - w \leq 0 \tag{8.32}$$
$$\pi_1 + \pi_2 - w \leq 100 \tag{8.33}$$
$$\pi_1 + \pi_2 + \pi_3 - w \leq 625 \tag{8.34}$$
$$\pi_1 + \pi_2 + \pi_4 - w \leq 800 \tag{8.35}$$
$$\pi_i \in \mathbb{R} \qquad\qquad i \in N$$
$$w \geq 0.$$

Without the pre-processing procedure which adds (8.32), we would need one iteration more which generates this constraint:

$$\pi_1 + \pi_4 - w \leq 800.$$

Note, the formulation of the objective function in the subproblem (8.21) provokes that those subcoalitions containing players not participating in the production process are found (and not the subcoalition without such players). This is due to the fact that the objective function value is higher while having players in the found coalition who might increase $\sum_{i \in N} \pi_i z_i$ but not o.

The solution above indicates that the algorithm also works for games with restricted cooperation and generates only some (four out of nine) of the rationality constraints based on the feasible coalitions.

As announced beforehand, we can show with the help of this example that MLCLSP games are not concave in general:

$$c(\{1, 2, 3\}) - c(\{1, 2\}) = 625 - 100 = 525$$
$$< 650 = 1,450 - 800 = c(N) - c(\{1, 2, 4\}).$$

The found core cost allocation does not seem to be very fair: In spite of having own external demand, player 1 gets paid for its production (negative costs). Players 3 and 4 have to share the total costs including the payment for player 1's external demand. This leads over to the next section where we will discuss using some already introduced core variants to compute core allocations that are observed as more fair.

8.2.3 Computing Core Variants

As the MLCLSP game is also a non-monotone game, the *subcoalition-perfect core* seems to be appropriate. This core variant assures that no coalition S has to carry higher costs in the grand coalition than in any other supercoalition $\hat{S} \supset S$ ($\hat{S} \subseteq N$). See Sect. 2.3.7 for a detailed discussion of the subcoalition-perfect core. The proof that the set of non-negative core allocations equals the subcoalition-perfect core $C^+(N, c) = C^{SP}(N, c)$ can also be applied to games with restricted cooperation. Obviously, the coalition S and supercoalition $\hat{S}$ have to be feasible coalitions (S, $\hat{S} \in \mathcal{F}$). This core variant prevents that any player will be subsidized which might happen while using the classical core.

We use the master problem (3.13)–(3.17) (see p. 50). We fix those π_i-values to zero where the players have no external demand and skip the stability constraints for $S \in \mathcal{S}$ because negative values for the π_i are not feasible in the setting of the subcoalition-perfect core. For the numerical example from the last section, the subcoalition-perfect core is empty ($v = 12.5$, $\pi_1 = \pi_2 = 0$, $\pi_3 = 637.5$ and $\pi_4 = 812.5$). Take the core constraints (8.31)–(8.35): Obviously, restriction (8.34) in combination with (8.31) implies $\pi_4 \geq 825$ and (8.35) with (8.31) implies $\pi_3 \geq 650$ which forces $\pi_1 < 0$ due to $\pi_2 = 0$. In coalition $\{1, 2, 3\}$, player 1 produces the pre-product for player 3 and, in coalition $\{1, 2, 4\}$, player 1 produces the pre-product for player 4. However, in the grand coalition, player 1 cannot produce for both players for lack of capacity. Due to that, total costs rise.

Apart from the fact of neglecting negative cost shares, we can utilize the *minmax core* to calculate fair cost allocations. Compare Sect. 2.3.6 for further details concerning the minmax core. The discussion about the consequences of coalitions with $c(S) = 0$ applies here: Total costs of zero for a feasible coalition may enforce $\eta^* > 1$. But this does not need to be necessarily true if the characteristic function is not monotone. The master problem for the minmax core is given by (3.8)–(3.12) (see p. 50).

As for the classical core, the set $\mathcal{S}$ contains all single coalitions where the player has no external demand.

When computing an element in the minmax core, the *subproblem* needs changes in the objective function and an additional restriction to take the restricted coalitions into account: $\hat{SP}(\pi)$:

$$\max -\eta o + \sum_{i \in N} \pi_i z_i$$

s.t. (8.22)–(8.30) and

$$\sum_{i \in N} z_i \leq |N| - 1.$$

As long as there are insufficient stability restrictions (at least $|N| - 1$) in the master problem, η takes the value zero and the subproblem might calculate $c(S)$ that are higher than the optimal solution for the corresponding subcoalition. Thereafter, restrictions with optimal $c(S)$-values are generated and the algorithm leads to a minimal η^* and the corresponding cost allocation for the minmax core. For the numerical example, the following master problem results from the last iteration with the cost allocation $\pi_1 = -49.152$, $\pi_2 = 0$, $\pi_3 = 663.559$, $\pi_4 = 835.593$, and $\eta^* = 0.983051$:

$$\min \eta$$

s.t.

$$\pi_1 + \pi_2 + \pi_3 + \pi_4 = 1{,}450$$
$$\pi_2 \leq 0$$
$$\pi_1 \leq 1{,}455\eta$$
$$\pi_1 + \pi_3 \leq 1{,}560\eta$$
$$\pi_2 + \pi_4 \leq 1{,}500\eta$$
$$\pi_1 + \pi_2 + \pi_3 \leq 625\eta$$
$$\pi_1 + \pi_4 \leq 800\eta$$
$$\pi_1 + \pi_3 \leq 625\eta$$
$$\pi_2 + \pi_4 \leq 850\eta$$
$$\pi_i \in \mathbb{R} \qquad\qquad\qquad i \in N$$
$$\eta \geq 0.$$

The solution shows that for this instance the minmax core generates a cost allocation that subsidizes player 1 even more than in the classical core. π_3 and π_4 could be to some extent arbitrarily high because there are no stability constraints like $\pi_3 \leq c(\{3\})$ or $\pi_4 \leq c(\{4\})$. However, this behavior depends on the specific instance. Thus, the minmax core is not bad in general. One possibility to avoid this subsidizing is to combine the ideas of the minmax core and the subcoalition-perfect core. We simply call it the *positive minmax core* $MP^{M+}(S)$. The master problem is nearly the same as for the minmax core (3.8)–(3.12) unless (3.11) which have to be replaced by

$$\pi_i \geq 0 \qquad\qquad\qquad i \in N.$$

Obviously, if the subcoalition-perfect core is empty, the optimal η for the positive minmax core will be equal or greater than 1.

8.3 Computational Study for the MLCLSP Game

We implemented the proposed algorithm with AMPL/CPLEX 10.0.0. The tests were run on Intel Pentium hardware with 2.8 GHz and 504 MB RAM. Instances with ten and twenty players were analyzed. In both cases, the players were arranged in four different network structures – one "long" and three "wide" variants. In

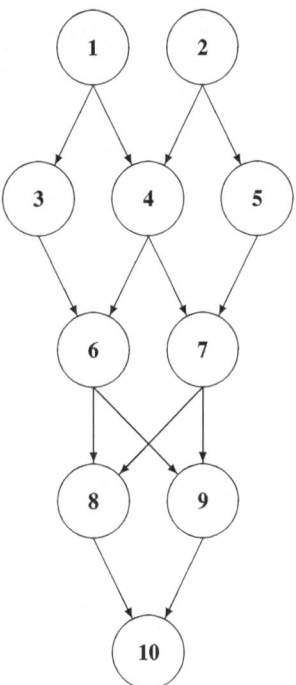

Fig. 8.3 MLCLSP game: supply network "long" with $|N| = 10$

the long network the players are organized at five (for $|N| = 10$) and nine (for $|N| = 20$) levels, respectively. In contrast, the wide networks consist of three (for $|N| = 10$) and four to five (for $|N| = 20$) levels, respectively. Compare Fig. 8.3 and Figs. B.1–B.7 in the Appendix for the detailed structure of the networks. The time horizon is $T = 6$ for all instances. For every type of network, 15 instances were generated where the remaining parameters are drawn from the following intervals with uniform distribution (with l as the number of levels in the network): $d_{it} \in [0; 20]$, $c_{it}^s \in [0; 200]$, $c_{it}^h \in [0; 10]$, $c_{it}^p \in [0; 15]$, $c_{ijt}^t \in [0; 15]$, $b_i \in [1; 5]$, $R_{it} \in [b_{it} * \sum_{s=t}^{T} d_{it}; l * b_i * \sum_{s=t}^{T} d_{it}]$.

Tables 8.2 and 8.3 display the results for the eight networks. Five of the instances from network "wide2" for $|N| = 10$ and "long" for $|N| = 20$ are infeasible because capacities for the suppliers are not enough to produce all the external demand in the grand coalition. These instances are excluded from the values in Tables 8.2 and 8.3. The numbers concerning the subcoalition-perfect core in Table 8.3 are based on feasible instances with a nonempty subcoalition-perfect core. The same is considered for the values regarding the positive minmax core. The results show that the number of iterations for long networks is lower than for the wide networks. This is induced by the structure of the long network where fewer coalitions are feasible compared to the wide networks.

Table 8.2 MLCLSP game: average results core and minmax core

$\lvert N \rvert$		$MP(\mathcal{S})$			$MP^M(\mathcal{S})$						# Feasible instances
		# Iterations			# Iterations			η			
		min	max	avg	min	max	avg	min	max	avg	
10	"long"	14	32	22	22	52	41	0.0000	0.9765	0.8951	15
	"wide1"	11	46	17	43	69	60	0.9950	1.0000	0.9989	15
	"wide2"	22	46	29	42	55	48	0.9765	1.0000	0.9905	10
	"wide3"	10	52	31	38	59	49	0.9739	1.0000	0.9901	15
20	"long"	64	710	223	128	334	232	0.0000	0.9885	0.7822	10
	"wide1"	141	596	346	156	718	363	0.9920	1.0000	0.9966	15
	"wide2"	184	578	368	124	236	177	0.9638	0.9895	0.9816	15
	"wide3"	53	591	329	229	646	394	0.9922	1.0000	0.9972	15

Table 8.3 MLCLSP game: average results subcoalition-perfect core and positive minmax core

$\lvert N \rvert$		$MP^+(\mathcal{S})$			$MP^{M+}(\mathcal{S})$						# Instances with $C^{scp} \neq \emptyset$
		# Iterations			# Iterations			η			
		min	max	avg	min	max	avg	min	max	avg	
10	"long"	19	45	32	30	43	35	0.9644	0.9964	0.9775	13
	"wide1"	11	54	22	48	66	56	0.9950	1.0000	0.9989	15
	"wide2"	26	48	36	23	58	39	0.9845	1.0000	0.9934	7
	"wide3"	10	42	32	31	43	37	0.9820	1.0000	0.9911	15
20	"long"	109	382	200	53	88	76	0.9767	0.9909	0.9844	9
	"wide1"	228	650	404	122	572	272	0.9936	1.0000	0.9969	13
	"wide2"	63	558	257	50	113	69	0.9834	0.9991	0.9902	10
	"wide3"	53	736	366	110	294	191	0.9922	1.0000	0.9974	15

We already saw in Sect. 7.3.3 that the row generation procedure generates more stability constraints for the minmax core than for the classical core in most of the cases. The same can be observed here (see Table 8.2). Adding the restriction for non-negative π_i results in a decreasing number of iterations for the positive minmax core compared to the minmax core. As expected, the optimal η for the positive minmax core is at least as high as the optimal η for the minmax core.

Remarkably, there are instances where the algorithm yields a core allocation in the minmax core with an optimal $\eta^* = 0$. One instance with ten players in the long network (see Fig. 8.3) with $c(N) = 25{,}272$ yields the following cost allocation for the minmax core

$$\pi_1 = \pi_2 = -101{,}088$$
$$\pi_5 = 50{,}544$$
$$\pi_3 = \pi_4 = \pi_6 = \pi_7 = \pi_8 = \pi_9 = \pi_{10} = 25{,}272.$$

Table 8.4 MLCLSP game: average results regarding computational time – average run-time (CPU-seconds)

$|N|$		$MP(\mathcal{S})$ CPU-seconds			$MP^M(\mathcal{S})$ CPU-seconds		
		min	max	avg	min	max	avg
10	"long"	0.45	1.05	0.71	0.66	1.72	1.37
	"wide1"	0.34	1.30	0.52	1.30	2.41	2.03
	"wide2"	0.66	1.47	0.97	1.19	1.80	1.50
	"wide3"	0.36	1.91	1.06	1.22	1.97	1.61
20	"long"	3.50	57.45	15.57	5.56	20.14	13.23
	"wide1"	8.30	41.69	22.34	8.09	49.72	21.86
	"wide2"	10.03	37.59	22.25	5.91	13.16	9.34
	"wide3"	2.56	47.81	21.72	11.83	43.73	22.53
$|N|$		$MP^+(\mathcal{S})$ CPU-seconds			$MP^{M+}(\mathcal{S})$ CPU-seconds		
		min	max	avg	min	max	avg
10	"long"	0.70	1.55	1.07	0.86	1.67	1.18
	"wide1"	0.31	1.83	0.71	1.48	2.47	1.99
	"wide2"	0.86	1.69	1.27	0.88	1.83	1.31
	"wide3"	0.36	1.61	1.13	1.00	1.47	1.21
20	"long"	6.23	26.77	12.55	2.61	5.55	4.48
	"wide1"	11.67	39.41	26.64	6.80	44.88	17.13
	"wide2"	3.63	31.66	14.57	2.64	5.58	3.43
	"wide3"	2.36	52.02	23.26	5.23	16.67	10.94

The reason for this strange result lies in the structure of the network and, hence, the structure of the feasible coalitions. All feasible coalitions (thus all stability constraints) have to contain players 1 and/or 2. Thus, it is possible to construct allocations where such "necessary" players have negative cost shares of an amount so that "essential" positive cost shares of other players meet all stability constraints. In the computational study, this happens only while using the long networks (one out of 15 instances for $|N| = 10$ and two out of ten instances for $|N| = 20$).

The computational time is displayed in Table 8.4. All instances can be solved in reasonable time and the CPU time per iteration is less than a second. Obviously, the computational time directly depends on the number of iterations.

Furthermore, we investigated instances where one or several players have no external demand. We take the network "wide3" and set demand to zero for a single player on the first level (1), for two players (1, 2), for all players (1, 2, 3), one player on the second level (4), two players with one for the first and another for the second level (1, 4), and all players on the first two levels (1, 2, 3, 4, 5, 6, 7). Due to our starting set $\mathcal{S}$ and the fixing pre-processing for the subcoalition-perfect core as well as the positive minmax core, the number of iterations should decrease. The results in Table 8.5 document this. The instances where only players on the third level in the supply network (player 8, 9 and 10) have external demand need a significantly

Table 8.5 MLCLSP game: average results for instances containing players without external demand

$\|N\| = 10$ "wide3" $\{i : \sum_{t=1}^{T} d_{it} = 0\}$	$MP(\mathcal{S})$ # Iterations			$MP^{M}(\mathcal{S})$ # Iterations			η			# Feasible instances
	min	max	avg	min	max	avg	min	max	avg	
1	9	40	23	40	58	48	0.9694	1.0000	0.9890	15
1, 2	9	21	17	33	60	48	0.9707	1.0000	0.9883	15
1, 2, 3	7	25	15	38	57	46	0.9683	0.9999	0.9889	15
4	11	32	20	38	76	52	0.9818	1.0000	0.9944	15
1, 4	8	24	17	35	66	47	0.9810	1.0000	0.9948	15
1, 2, 3, 4, 5, 6, 7	3	6	3	11	27	15	0.9931	1.0000	0.9990	15

$\|N\| = 10$ "wide3" $\{i : \sum_{t=1}^{T} d_{it} = 0\}$	$MP^{+}(\mathcal{S})$ # Iterations			$MP^{M+}(\mathcal{S})$ # Iterations			η			# Instances with $C^{scp} \neq \emptyset$
	min	max	avg	min	max	avg	min	max	avg	
1	9	49	29	29	43	38	0.9781	1.0000	0.9917	14
1, 2	10	29	18	30	38	33	0.9784	1.0000	0.9915	10
1, 2, 3	7	23	14	25	33	29	0.9794	0.9998	0.9939	9
4	11	35	21	23	53	38	0.9818	1.0000	0.9944	15
1, 4	8	25	18	25	43	33	0.9810	1.0000	0.9952	15
1, 2, 3, 4, 5, 6, 7	3	3	3	9	11	9	0.9931	1.0000	0.9989	14

smaller number of iterations. The higher number of stability constraints fixed before starting the procedure might have an influence on whether or not an element in the (subcoalition-perfect) core can be found. A bigger number of instances with an empty subcoalition-perfect core appeared when players on the first level had no demand (in particular for $\{1, 2\}$ and $\{1, 2, 3\}$).

Chapter 9
Conclusions and Future Research

The present work has brought together two fields of research – cooperative game theory and lot sizing in supply chains – to heighten the awareness for practical problems occurring in supply networks where several partners are involved in a cooperation and to show the applicability of cooperative game theory to practical problems. Although, many questions arise when dealing with cooperations, this work has concentrated on the problem of allocating cooperative profits or costs among the partners of a cooperation because this plays an essential role to assure stability and fairness in a cooperation.

We started in Chap. 2 with presenting selected topics from cooperative game theory to give an abstract mathematical description of how to handle cooperative decision making. Additionally, we discussed important properties of cooperative games (like monotonicity, subadditivity, and concavity) and gave an overview of game variants and fundamental applications that are studied in the literature. Afterwards, we discussed one of the most prominent concepts for allocation: the core. This concept combines two fundamental properties an allocation should have: The first property is efficiency; i.e., total costs have to be allocated completely (neither more nor less), and the second is called stability; i.e., a coalition or single player should not bear more costs than when acting alone without the rest of the grand coalition. Based on critical points regarding the core concept, we have discussed well known variants of the core. In this discussion, we concentrated on refining and extending two known core variants, the minmax core and the interval core. Furthermore, we developed a new core variant, the subcoalition-perfect core, that should be used particularly for non-monotone cooperative games which has not been investigated in the literature up to now. We showed that for cost games the subcoalition-perfect core coincides with the set of non-negative core allocations.

We chose the core and its variants for our further study because these concepts assure stability of a cooperation – it is essential for the reliability of a cooperation that no player has an incentive to leave the cooperation. Simultaneously, the core offers flexibility to bring in aspects of fairness which means, for instance, to control the allocated shares in a way that profit or cost shares are equally distributed among the cooperating partners.

When applying the core or its related concepts to complex practical decision problems, we need an algorithm that is able to compute core elements sufficiently

J. Drechsel, *Cooperative Lot Sizing Games in Supply Chains*, Lecture Notes
in Economics and Mathematical Systems 644, DOI 10.1007/978-3-642-13725-9_9,
© Springer-Verlag Berlin Heidelberg 2010

fast. We provided such an algorithm in Chap. 3. Describing the core needs plenty of stability constraints but not all of them are essentially needed to describe the core or denote a core allocation, respectively. Hence, the procedure is based on the idea of row (or constraint) generation: We start the computation with only a few stability constraints and add a new stability constraint iteration by iteration until a core allocation is found. This procedure is general because it is applicable to a large class of cooperative optimization problems. Furthermore, the algorithm is not only flexible in its application to concrete problems but it can be easily adapted to compute elements for the presented core variants.

Following this theoretical basis, we discussed the importance of cooperation, in particular, in the field of modern supply chain management (see Chap. 4). Thereby, we distinguished between horizontal and vertical cooperations in supply networks. As one example, we introduced purchasing alliances that are well established in praxis and that are the basis for the first mathematical formulation of a cooperative ordering situation in Chap. 5.

Beginning with Chap. 5, we combined the two streams; i.e., the developed algorithm to compute fair cost allocations is applied to concrete cooperative problems. We started with the cooperative economic lot sizing problem (ELS game) where partners cooperate to make joint orders. The main incentive for joint ordering is to reduce fixed ordering costs that can be shared among the partners of the cooperation. The cooperative ELS is already discussed in the literature but without providing solution algorithms. We showed with an extensive computational study that our proposed row generation algorithm is very efficient when computing core elements; i.e., a very small percentage of the core describing stability constraints is generated until a core element is found. In the following chapters, we developed more complex cooperative lot sizing problems that are not part of the research literature up to now.

In Chap. 6, we expanded the ELS game to handle uncertain demand. The possibility of using intervals to represent uncertainty is not new, but its application to a concrete problem like uncertain demand in ordering situations. Authors dealing with interval-valued games remark that computing interval core elements is easy because it is sufficient to determine a core element for the lower interval and another for the upper interval separately. However, we revealed that there are some interpretation problems of the interval core when computing the core as proposed; i.e., cost shares for the lower interval might be greater than for the upper interval. Therefore, we developed a variation of the row generation procedure by adjusting the master problem to compute upper and lower interval core elements simultaneously.

Chapter 7 provided the formulation of a more complex lot sizing problem that included cooperative production, transshipments among the cooperating partners, scarce production capacities, and player-dependent cost coefficients. This means players in a cooperation have limited capacities for their production but can pool the capacities for a better utilization. Due to these new restrictions, it is a hard to solve problem when computing optimal costs for the grand coalition. Hence, we developed two heuristical methods (a Lagrangean relaxation based heuristic and a fix-and-optimize heuristic) to determine optimal costs for the grand coalition and tested them in computational studies. Furthermore, we investigated that this game has the specialty of being not monotone; i.e., the total costs for a coalition may decrease with a bigger size. Therefore, we applied the subcoalition-perfect core and showed with a computational study that the row generation procedure can efficiently compute elements in the subcoalition-perfect core.

The setting of cooperative production is furthermore extended to multilevel structures in Chap. 8. This extension causes the consideration of restricted cooperation because now not every arbitrary subcoalition is feasible; e.g., in multilevel structures a minimum number of pre-products is needed to produce a special product. The world of restricted cooperation is not new in the literature, but is studied mostly from a theoretical point of view where only very special structures of the feasible coalitions are taken into account. None of these structures could be applied to our multilevel supply structure. With a computational study, we showed that our row generation algorithm can handle cooperative games with restricted cooperation very efficiently as well.

The overall contribution of this work is that it provides a theoretical basis when dealing with allocation problems in any kind of cooperation and, furthermore, possible applications in the field of supply chain management. The investigated cooperative lot sizing problems are not only challenging optimization problems but also challenging for cooperative game theory because none of those games is a concave cost game and, furthermore, when including limited production capacities, transshipment cost coefficients, and/or player dependent cost coefficients, the games are not monotone. Such games are disregarded mostly in the literature up to now and we present concepts to handle them.

The presented cooperative lot sizing problems can be used like a construction kit to develop further and more specialized cooperative lot sizing problems; e.g., it might be of interest to formulate problems where players are not business units or companies but products or orders. As already discussed, the row generation procedure to compute core elements is general and might be applied to many other applications – not only in cooperative lot sizing; e.g., to problems of cooperative vehicle routing. Additionally, it might be of interest to analyze how sensitive the computed core allocation reacts to parameter changes. Another field for potential research may be to study information transactions in a cooperation to minimize the size of information that need to be shared in a coalition. Furthermore, the application of the nucleolus to the described problems of cooperative lot sizing waits for examination. But first of all, there are many research opportunities in investigating cooperative cost games that are not monotone and not concave because it seems to be that many practical cooperative problems do not have those helpful properties.

Appendix A
Computational Study CLSP Game

A.1 Computational Study: Lagrangean Relaxation Based Heuristic

Table A.1 CLSP game: average results Lagrangean relaxation based heuristic for a varying number of products

	$|N|$	$|K|$	$[\underline{c}^l_{ijkt}; \overline{c}^l_{ijkt}]$	Avg %-Gap(LB)	Avg %-Gap(UB)	Avg # iterations	Avg CPU time [s]
h1	4	3	[0;0]	14.61%	0.01%	238.93	26.20
	4	3	[0;5]	8.46%	0.03%	255.20	29.11
	4	3	[0;15]	4.13%	0.04%	242.00	27.49
	4	3	[0;50]	3.30%	0.15%	237.93	26.48
	4	3	[5000;5000]	4.19%	0.13%	255.53	28.29
	4	15	[0;0]	1.98%	0.11%	255.07	138.03
	4	15	[0;5]	0.98%	0.08%	282.33	153.62
	4	15	[0;15]	0.38%	0.06%	282.00	149.90
	4	15	[0;50]	0.29%	0.06%	388.20	204.36
	4	15	[5000;5000]	0.73%	0.05%	545.27	302.06
h3	4	3	[0;0]	14.64%	0.48%	229.73	18.67
	4	3	[0;5]	8.46%	0.34%	257.07	21.05
	4	3	[0;15]	4.13%	0.26%	241.47	19.50
	4	3	[0;50]	3.31%	0.32%	237.07	19.25
	4	3	[5000;5000]	4.22%	0.36%	264.47	21.28
	4	15	[0;0]	1.97%	0.28%	262.13	120.87
	4	15	[0;5]	0.97%	0.15%	250.73	115.40
	4	15	[0;15]	0.37%	0.11%	254.73	117.97
	4	15	[0;50]	0.28%	0.07%	333.00	153.54
	4	15	[5000;5000]	0.73%	0.11%	441.13	222.21
h6	4	3	[0;0]	14.66%	9.53%	250.73	26.01
	4	3	[0;5]	8.49%	12.80%	255.73	26.93
	4	3	[0;15]	4.13%	19.90%	236.00	24.54
	4	3	[0;50]	3.22%	17.58%	235.00	24.64
	4	3	[5000;5000]	4.17%	10.29%	226.80	23.44
	4	15	[0;0]	1.94%	4.94%	254.60	138.21
	4	15	[0;5]	0.94%	13.66%	236.33	131.12
	4	15	[0;15]	0.36%	20.04%	231.27	127.23
	4	15	[0;50]	0.27%	18.33%	226.47	123.51
	4	15	[5000;5000]	0.64%	11.42%	266.87	145.27

Table A.2 CLSP game: average results Lagrangean relaxation based heuristic variant h1 for varying transportation cost coefficients and for a varying number of players

$\lvert N \rvert$	$\lvert K \rvert$	$[\underline{c}^t_{ijkt}; \overline{c}^t_{ijkt}]$	Avg %-Gap(LB)	Avg %-Gap(UB)	Avg # iterations	Avg CPU time [s]
3	1	[0;0]	35.63%	0.00%	242.20	12.71
3	1	[0;5]	27.96%	0.00%	248.93	14.19
3	1	[0;15]	20.30%	0.00%	237.40	13.15
3	1	[0;50]	14.40%	0.00%	231.07	12.53
3	1	[5000;5000]	13.30%	0.00%	227.27	11.95
3	3	[0;0]	11.44%	0.04%	242.93	25.79
3	3	[0;5]	6.84%	0.07%	229.60	24.72
3	3	[0;15]	3.51%	0.04%	221.60	23.45
3	3	[0;50]	3.28%	0.01%	253.53	26.34
3	3	[5000;5000]	5.78%	0.10%	245.40	25.64
5	3	[0;0]	16.18%	0.03%	264.20	33.60
5	3	[0;5]	9.67%	0.00%	285.00	39.05
5	3	[0;15]	5.08%	0.05%	255.80	34.93
5	3	[0;50]	3.26%	0.06%	250.80	32.44
5	3	[5000;5000]	5.09%	0.03%	263.67	33.00
10	3	[0;0]	27.81%	0.00%	314.67	51.37
10	3	[0;5]	17.30%	0.00%	311.33	63.49
10	3	[0;15]	7.70%	0.02%	295.27	65.06
10	3	[0;50]	3.21%	0.10%	291.33	56.65
10	3	[5000;5000]	4.57%	0.10%	275.73	43.96
15	3	[0;0]	32.52%	0.00%	341.80	70.23
15	3	[0;5]	24.26%	0.00%	361.13	89.31
15	3	[0;15]	11.25%	0.04%	307.00	102.78
15	3	[0;50]	4.17%	0.08%	293.80	79.39
15	3	[5000;5000]	4.86%	0.13%	302.00	68.74

Table A.3 CLSP game: average results Lagrangean relaxation based heuristic variant h3 for varying transportation cost coefficients and for a varying number of players

$|N|$	$|K|$	$[\underline{c}^t_{ijkt}; \overline{c}^t_{ijkt}]$	Avg %-Gap(LB)	Avg %-Gap(UB)	Avg # iterations	Avg CPU time [s]
3	1	[0;0]	35.69%	0.12%	240.67	6.75
3	1	[0;5]	27.94%	0.00%	247.33	7.52
3	1	[0;15]	20.30%	0.00%	240.00	7.20
3	1	[0;50]	14.39%	0.07%	227.93	6.70
3	1	[5000;5000]	13.23%	0.09%	220.27	6.10
3	3	[0;0]	11.48%	0.36%	239.07	18.59
3	3	[0;5]	6.86%	0.18%	236.93	18.30
3	3	[0;15]	3.50%	0.27%	226.00	17.60
3	3	[0;50]	3.30%	0.26%	256.87	20.06
3	3	[5000;5000]	5.79%	0.22%	241.20	18.89
5	3	[0;0]	16.81%	0.30%	267.13	23.74
5	3	[0;5]	9.69%	0.34%	251.87	23.65
5	3	[0;15]	5.08%	0.22%	266.87	24.44
5	3	[0;50]	3.24%	0.31%	270.13	25.73
5	3	[5000;5000]	5.11%	0.23%	271.47	24.70
10	3	[0;0]	27.77%	0.23%	309.93	32.35
10	3	[0;5]	17.33%	0.17%	296.87	32.73
10	3	[0;15]	7.72%	0.16%	289.60	33.63
10	3	[0;50]	3.22%	0.31%	286.13	30.48
10	3	[5000;5000]	4.54%	0.36%	275.13	30.47
15	3	[0;0]	32.45%	0.20%	352.67	48.06
15	3	[0;5]	24.20%	0.14%	343.67	55.53
15	3	[0;15]	11.24%	0.21%	311.13	75.54
15	3	[0;50]	4.19%	0.32%	289.93	51.35
15	3	[5000;5000]	4.86%	0.31%	273.67	43.10

Table A.4 CLSP game: average results Lagrangean relaxation based heuristic variant h6 for varying transportation cost coefficients and for a varying number of players

$|N|$	$|K|$	$[\underline{c}_{ijkt}^t; \overline{c}_{ijkt}^t]$	Avg %-Gap(LB)	Avg %-Gap(UB)	Avg # iterations	Avg CPU time [s]
3	1	[0;0]	35.95%	6.46%	253.87	12.60
3	1	[0;5]	28.18%	8.05%	259.47	13.59
3	1	[0;15]	20.30%	8.61%	238.00	12.16
3	1	[0;50]	14.38%	7.69%	235.07	12.10
3	1	[5000;5000]	13.18%	5.04%	240.93	11.85
3	3	[0;0]	11.48%	9.99%	231.40	23.20
3	3	[0;5]	6.87%	13.89%	231.13	23.28
3	3	[0;15]	3.47%	17.71%	223.00	22.24
3	3	[0;50]	3.26%	15.42%	227.80	22.69
3	3	[5000;5000]	5.67%	8.66%	241.60	23.95
5	3	[0;0]	16.82%	10.79%	264.93	30.50
5	3	[0;5]	9.62%	15.03%	262.87	30.83
5	3	[0;15]	4.99%	21.40%	257.87	30.55
5	3	[0;50]	3.21%	20.23%	253.47	29.69
5	3	[5000;5000]	5.00%	10.75%	242.00	27.84
10	3	[0;0]	27.91%	11.61%	310.73	38.17
10	3	[0;5]	17.38%	15.68%	310.00	42.05
10	3	[0;15]	7.75%	20.88%	290.20	41.83
10	3	[0;50]	3.20%	26.07%	269.80	36.75
10	3	[5000;5000]	4.45%	11.97%	254.00	33.86
15	3	[0;0]	32.74%	12.36%	355.33	58.13
15	3	[0;5]	24.44%	14.31%	344.80	66.74
15	3	[0;15]	11.26%	20.39%	306.60	82.67
15	3	[0;50]	4.14%	25.12%	280.80	62.19
15	3	[5000;5000]	4.75%	11.28%	273.53	51.11

A.2   Computational Study: Fix-and-Optimize Heuristic

Table A.5 CLSP Game: results for fix-and-optimize heuristic for varying transportation cost coefficients and a varying number of players (plb+prb+tb, * response time more than 3 h)

$[\underline{c}_{ijkt}^{t}; \overline{c}_{ijkt}^{t}]$	$\|N\|$	$\|K\|$	Avg %- Gap(UB)	Avg # iterations	Avg CPU time [s]	$\|N\|$	$\|K\|$	Avg %- Gap(UB)	Avg # iterations	Avg CPU time [s]
[0; 0]	3	1	0.00%	2.00	0.30	15	3	0.13%	2.73	1.57
[0; 5]	3	1	0.00%	2.00	0.26	15	3	0.02%	3.27	2.22
[0; 15]	3	1	0.00%	2.00	0.27	15	3	0.04%	2.80	1.92
[0; 50]	3	1	0.00%	2.00	0.28	15	3	0.00%	2.87	2.11
[5000; 5000]	3	1	0.00%	1.87	0.26	15	3	0.00%	2.00	1.91
[0; 0]	3	3	0.14%	2.20	0.39	20	3	0.07%	2.67	1.90
[0; 5]	3	3	0.04%	2.40	0.47	20	3	0.04%	3.13	2.79
[0; 15]	3	3	0.21%	2.47	0.48	20	3	0.03%	2.73	2.54
[0; 50]	3	3	0.02%	2.33	0.48	20	3	0.00%	2.87	2.92
[5000; 5000]	3	3	0.00%	2.00	0.37	20	3	0.00%	2.00	37.99
[0; 0]	4	15	0.08%	2.87	1.43	25	3	0.06%	2.87	2.66
[0; 5]	4	15	0.04%	3.33	2.09	25	3	0.04%	2.67	2.93
[0; 15]	4	15	0.02%	3.13	1.94	25	3	0.01%	3.13	3.89
[0; 50]	4	15	0.02%	2.93	1.92	25	3	0.01%	3.07	4.56
[5000; 5000]	4	15	0.00%	2.00	1.33	25	3	0.00%*	2.20*	140.07*
[0; 0]	5	3	0.08%	2.53	0.53	50	3	0.03%	3.40	13.19
[0; 5]	5	3	0.01%	2.60	0.64	50	3	0.04%	3.20	14.51
[0; 15]	5	3	0.01%	2.33	0.58	50	3	0.01%	3.47	19.31
[0; 50]	5	3	0.01%	2.27	0.65	50	3	0.00%	3.33	30.15
[5000; 5000]	5	3	0.00%	2.00	0.50	50	3	–*	–*	–*
[0; 0]	10	3	0.11%	2.47	0.86	100	3	0.02%	3.53	125.09
[0; 5]	10	3	0.04%	2.60	1.12	100	3	0.02%	3.20	135.98
[0; 15]	10	3	0.02%	2.40	1.06	100	3	0.02%	3.73	284.98
[0; 50]	10	3	0.00%	2.47	1.17	100	3	–*	–*	–*
[5000; 5000]	10	3	0.00%	2.07	1.02	100	3	–*	–*	–*

Appendix B
Computational Study MLCLSP Game

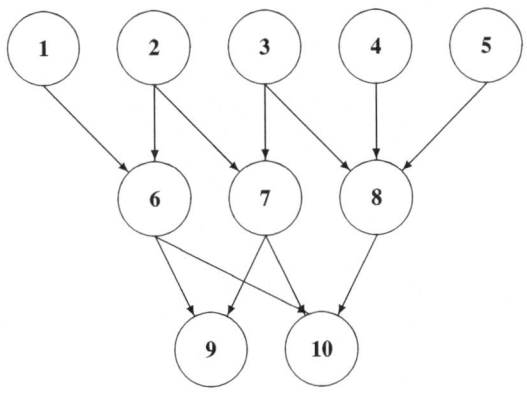

Fig. B.1 MLCLSP game: supply network "wide1" with $|N| = 10$

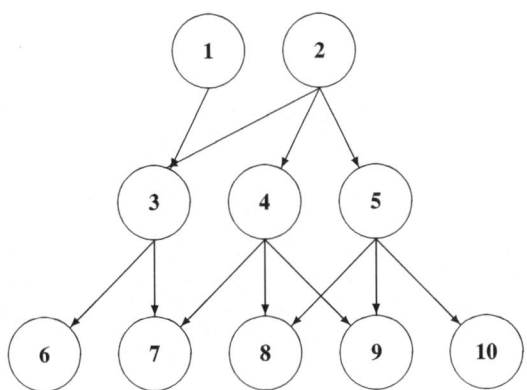

Fig. B.2 MLCLSP game: supply network "wide2" with $|N| = 10$

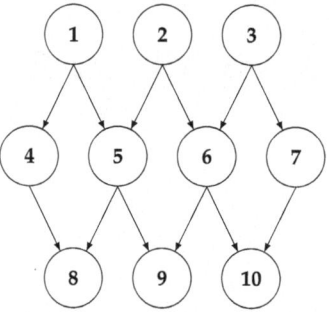

Fig. B.3 MLCLSP game: supply network "wide3" with $|N| = 10$

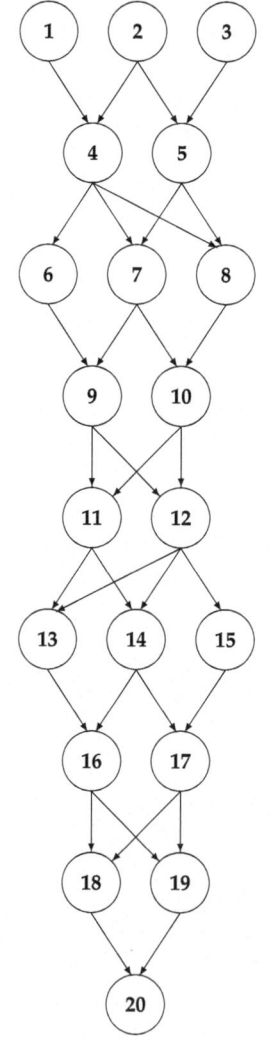

Fig. B.4 MLCLSP game: supply network "long" with $|N| = 20$

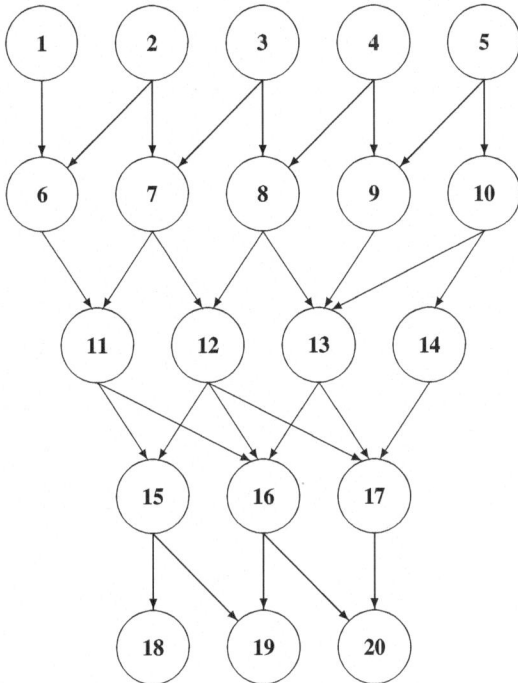

Fig. B.5 MLCLSP game: supply network "wide1" with $|N| = 20$

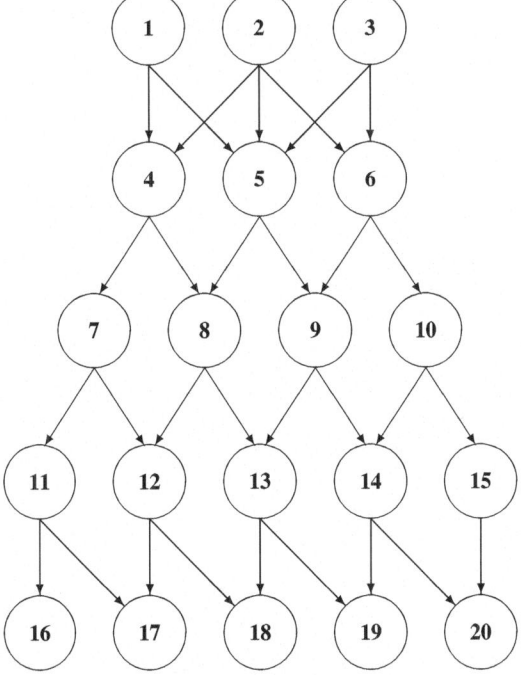

Fig. B.6 MLCLSP game: supply network "wide2" with $|N| = 20$

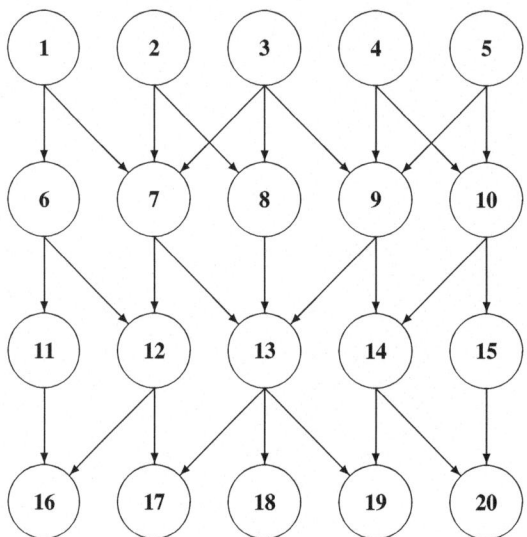

Fig. B.7 MLCLSP game: supply network "wide3" with $|N| = 20$

Bibliography

Aggarwal, A., & Park, J. K. (1993). Improved Algorithms for Economic Lot Size Problems. *Operations Research, 41*(3), 549–571.

Albizuri, M. J., Santos, J. C., & Zarzuelo, J. M. (2003). On the Serial Cost Sharing Rule. *International Journal of Game Theory, 31*(3), 437–446.

Algaba, E., Bilbao, J. M., van den Brink, R., & Jiménez-Losada, A. (2004). Cooperative Games on Antimatroids. *Discrete Mathematics, 282*(1-3), 1–15.

Algaba, E., Bilbao, J. M., & López, J. J. (2001) A Unified Approach to Restricted Games. *Theory and Decision, 50*(4), 333–345.

Alparslan-Gök, S. Z., Branzei, R., & Tijs, S. H. (2008a). *Cores and Stable Sets for Interval-Valued Games*. Working Paper Tilburg University.

Alparslan-Gök, S. Z., Branzei, R., & Tijs, S. H. (2008b). *Convex Interval Games*. Working Paper Tilburg University.

Alparslan-Gök, S. Z., Branzei, R., & Tijs, S. H. (2008c). *Big Boss Interval Games*. Working Paper Tilburg University.

Alparslan-Gök, S. Z., Branzei, R., & Tijs, S. H. (2008d). *Cooperative Interval Games Arising from Airport Situations with Interval Data*. Working Paper Tilburg University.

Alparslan-Gök, S. Z., Miquel, S., Tijs, S. H.(2009). Cooperation under Interval Uncertainty. *Mathematical Methods of Operations Research, 69*(1), 99–109.

Arcelus, F. J., Bhadury, J., Srinivasan, G. (1997). On the Interaction between Indirect Cost Allocations and the Firm's Objectives. *European Journal of Operational Research, 102*(3), 445–454.

Arnold, U. (1998). Einkaufskooperationen im Mittelstand. In R. Large, (Ed.), *Trends im Beschaffungsmanagement* pp. 31–50. Bernburg.

Arnold, U., & Eßig, M. (1997). *Einkaufskooperationen in der Industrie*. Schäffer-Poeschel Verlag, Stuttgart.

Aumann, R. J., & Maschler, M. (1964). The Bargaining Set for Coopertive Games, In M. Dresher, L. S. Shapley, A. W. Tucker (Eds.), *Advances in Game Theory, Annals of Mathematical Studies No. 52* pp. 443–476. Princeton University Press, Princeton.

Axsäter, S. (2003). Evaluation of Unidirectional Lateral Transshipments and Substitutions in Inventory Systems. *European Journal of Operational Research, 149*(2), 438–447.

Balachandran, B. V., & Ramakrishnan, R. T. S. (1981). Joint Cost Allocation: A Unified Approach. *The Accounting Review, 56*(1), 85–96.

Barnhart, C., Johnson, E. L., Nemhauser, G. L., Savelsbergh, M. W. P., & Vance, P. H. (1998). Branch-and-Price: Column Generation for Solving Huge Integer Programs. *Operations Research, 46*(3), 316–329.

Barton, T. L. (1988). Intuitive Choice of Cooperative Sharing Mechanisms for Joint Cost Savings: Some Empirical Results. *Abacus, 24*(2), 162–169.

Bauso, D., & Timmer, J. (2009). Robust Dynamic Cooperative Games. *International Journal of Game Theory, 38*(1), 23–36.

Benders, J. F. (1962). Partitioning Procedures for Solving Mixed-Variables Programming Problems. *Numerische Mathematik*, *4*(1), 238–252.

Bertrand, J. (1883). Review of Walras' 'Théorie Mathématique de la Richesse Sociale' and Cournot's 'Recherches sur les Principes Mathématiques de la Théorie des Richesses'. *Journal des Savants*, 499–508.

Biddle, G. C., & Steinberg, R. (1984). Allocations of Joint and Common Costs. *Journal of Accounting Literature*, *3*(1), 1–45.

Bilbao, J. M. (2000). *Cooperative Games on Combinatorial Structures*. Kluwer Academic Publishers.

Bilbao, J. M. (2003). Cooperative Games under Augmenting Systems. *SIAM Journal of Discrete Mathematics*, *17*(1), 122–133.

Bilbao, J. M., Fernández, J. R., Jiménez, N., & López, J. J. (2007). The Core and the Weber Set for Bicooperative Games. *International Journal of Game Theory*, *36*(2), 209–222.

Bilbao, J. M., Fernández, J. R., Jiménez, N., & López, J. J. (2008). The Shapley Value for Bicooperative Games. *Annals of Operations Research*, *158*(1), 99–115.

Bilbao, J. M., Jiménez, N., & Lebrón, E. (1999). The Core of Games on Convex Geometries. *European Journal of Operational Research*, *119*(2), 365–372.

Bilbao, J. M., Jiménez, N., Lebrón, E., & López, J. J. (2006). The Marginal Operators for Games on Convex Geometries. *International Game Theory Review*, *8*(1), 141–151.

Bilbao, J. M., & Martínez-Legaz, J. E. (2002). Some Applications of Convex Analysis to Cooperative Game Theory. *Journal of Statistics and Management Systems*, *5*(1), 39–61.

Bilbao, J. M., & Ordóñez, M. (2009). Axiomatizations of the Shapley Value for Games on Augmenting Systems. *European Journal of Operational Research*, *196*(3), 1008–1014.

Billington, P. J., McClain, J. O., & Thomas, L. J. (1983). Mathematical Programming Approaches to Capacity-Constrained MRP Systems: Review, Formulation and Problem Reduction. *Management Science*, *29*(10), 1126–1141.

Binmore, K. (1992). *Fun and Games – A Text on Game Theory*. D. C. Heath and Company, Lexington Massachusetts.

Bird, C. G. (1976). On Cost Allocation for a Spanning Tree: A Game Theoretic Approach. *Networks*, *6*(4), 335–350.

Bjørndal, E., & Jörnsten, K. (2009). Lower and Upper Bounds for Linear Production Games. *European Journal of Operational Research*, *196*(2), 476–486.

Bondareva, O. N. (1963). Some Applications of Linear Programming Methods to the Theory of Cooperative Games. *Problemy Kibernetiki*, 10, 119–139, In L. Billera (Ed.), *Selected Russian Papers on Game Theory 1959-1965*. Princeton University Press, Princeton (1968).

Bonnans, F., & André, M. (2008). Fast Computation of the Least Core and Prenucleolus of Cooperative Games. *RAIRO Operations Research*, *42*(3), 299–314.

Borm, P., Hamers, H., & Hendrickx, R. (2001). Operations Research Games: A Survey. *TOP, Sociedad de Estadística e Investigación Operativa*, *9*(2), 139–216.

Borm, P., Owen, G., & Tijs, S. H. (1992). On the Position Value for Communcation Situations. *SIAM Journal on Discrete Mathematics*, *5*(3), 305–320.

Bradley, S. P., Hax, C., Arnoldo, Magnanti, T. L. (1977). *Applied Mathematical Programming*. Addison-Wesley, Reading.

Brandenburger, A. M., & Harborne, W. S. J. (1996). Value-Based Business Strategy. *Journal of Economics and Management Strategy*, *5*(1), 5–24.

Brandenburger, A. M., & Nalebuff, B. J. (1996). *Co-opetition*. Doubleday, New York.

Branzei, R., & Alparslan-Gök, S. Z. (2008). Bankruptcy Problems with Interval Uncertainty. *Economics Bulletin*, *3*(56), 1–10.

Branzei, R., Dimitrov, D., Pickl, S., & Tijs, S. (2004). How to Cope with Division Problems under Interval Uncertainty of Claims. *International Journal of Uncertainty, Fuzziness and Knowledge-Based Systems*, *12*(2), 191–200.

Branzei, R., Dimitrov, D., & Tijs, S. (2003). Shapley-Like Values for Interval Brankruptcy Games. *Economics Bulletin*, *3*(9), 1–8.

Branzei, R., Dimitrov, D., & Tijs, S. (2008a). *Models in Cooperative Game Theory* (2nd ed.). Springer, Berlin.

Branzei, R., Tijs, S. H., & Alparslan-Gök, S. Z. (2008b). Some Characterizations of Convex Interval Games. *AUCO Czech Economic Review, 2*(3), 219–226.

Branzei, R., Tijs, S. H., & Alparslan-Gök, S. Z. (2008c). How to Handle Interval Solutions for Cooperative Interval Games. Working Paper Tilburg University.

Branzei, R., Tijs, S. H., & Alparslan-Gök, S. Z. (2009). Cooperative Interval Games: A Survey. *Central European Journal of Operations Research*, online appearance December 2009.

van den Brink, R. (1997). An Axiomatization of the Disjunctive Permission Value for Games with a Permission Structure. *International Journal of Game Theory, 26*(1), 27–43.

Buckley, P. J., & Casson, M. (1988). A Theory of Cooperation in International Business. In F. J. Contractor, P. Lorange (Eds.), *Cooperative Strategies in International Business* pp. 31–53. Lexington Books, Lexington.

Buschkühl, L., Sahling, F., Helber, S., & Tempelmeier, H. (2010). Dynamic Capacitated Lot-Sizing Problems: A Classification and Review of Solution Approaches. *OR Spectrum, 32*(2), 231–261.

Butnariu, D., & Klement, E. P. (1993). *Triangular Norm-Based Measures and Games with Fuzzy Coalitions*. Kluwer Academic Publishers.

Cachon, G. P., & Netessine, S. (2004). Game Theory in Supply Chain Analysis. In D. Simchi-Levi, S. D. Wu, Z. M. Shen (Eds.), *Handbook of Quantitative Supply Chain Analysis: Modeling in the E-Business Era* pp. 13–65. Springer.

Calvo, E., & Santos, J. C. (2000). A Value for Multichoice Games. *Mathematical Social Sciences, 40*(3), 341–354.

Carpente, L., Casas-Méndez, B., García-Jurado, I., & van den Nouweland, A. (2005). *Interval Values for Strategic Games in Which Players Cooperate*. Working Paper University of Oregon.

Çetiner, D., & Kimms, A. (2009). *Approximate Nucleolus Allocation in Airline Alliances*. Working Paper University Duisburg-Essen.

Chardaire, P. (2001). The Core and the Nucleolus of Games: A Note on a Paper by Göthe-Lundgren et al.. *Mathematical Programming, Series A, 90*(1), 147–151.

Chatterjee, K., & Samuelson, W. F. (Eds.). (2001). *Game Theory and Business Applications*. Kluwer Academic Publishers, Boston.

Chen, X. (2009). Inventory Centralization Games with Price-Dependent Demand and Quantity Discount. *Operations Research, 57*(6), 1394–1406.

Chen, X., & Zhang, J. (2006). Duality Approaches to Economic Lot-Sizing Games. Working Paper University of Illinois at Urbana-Champaign.

Chen, X., & Zhang, J. (2007). *A Stochastic Programming Duality Approach to Inventory Centralization Games*. Working Paper University of Illinois at Urbana-Champaign.

Claus, A., & Kleitman, D. J. (1973). Cost Allocation for a Spanning Tree. *Networks, 3*(4), 289–304.

Contractor, F. J., & Lorange, P., (Eds.). (1988a). *Cooperative Strategies in International Business*, Lexington Books, Lexington.

Contractor, F. J., & Lorange, P. (1988b). Why Should Firms Cooperate? The Strategy and Economic Basis for Cooperative Ventures. In F. J. Contractor, P. Lorange (Eds.), *Cooperative Strategies in International Business* pp. 3–30. Lexington Books, Lexington.

Cournot, A. (1838). *Recherches sur les Principes Mathématiques de la Théorie des Richesses*. Hachette, Paris.

Cruijssen, F., Cools, M., & Dullaert, W. (2007). Horizontal Cooperation in Logistics: Opportunities and Impediments. *Transportation Research Part E, 43*(2), 129–142.

Curiel, I. (1997). *Cooperative Game Theory and Applications*. Kluwer Academic Publishers, Boston.

Curiel, I., Pederzoli, G., & Tijs, S. (1989). Sequencing Games. *European Journal of Operational Research, 40*(3), 344–351.

Dantzig, G. B., & Wolfe, P. (1960). Decomposition Principle for Linear Programs. *Operations Research. 8*(1), 101–111.

Davis, M., & Maschler, M. (1965). The Kernel of a Cooperative Game. *Naval Research Logistics Quarterly, 12*(3/4), 223–259.

Demski, J. S. (1981). Cost Allocation Games. In S. Moriarity (ed.), *Joint Cost Allocations –*
Proceedings of the University of Oklahoma, Conference on Cost Allocation 142–173.

Deng, X., Ibaraki, T., & Nagamochi, H. (1999). Algorithmic Aspects of the Core of Combinatorial
Optimization Games. *Mathematics of Operations Research*, *24*(3), 751–766.

Deng, X., Ibaraki, T., Nagamochi, H., & Zang, W. (2000). Totally Balanced Combinatorial
Optimization Games. *Mathematical Programming*, *87*(3), 441–452.

Deng, X., & Papadimitriou, C. H. (1994). On the Complexity of Cooperative Solution Concepts.
Mathematics of Operations Research, *19*(2), 257–266.

Derks, J. J. M., & Gilles, R. P. (1995). Hierarchical Organization Structures and Constraints on
Coalition Formation. *International Journal of Game Theory*, *24*(2), 147–163.

Derks, J. J. M., & Kuipers, J. (1997). On the Core of Routing Games. *International Journal of*
Game Theory, *26*(2), 193–205.

Derks, J. J. M., & Peters, H. (1993). A Shapley Value for Games with Restricted Coalitions.
International Journal of Game Theory, *21*(4), 351–360.

Derks, J. J. M., & Reijnierse, H. (1998). On the Core of a Collection of Coalitions. *International*
Journal of Game Theory, *27*(3), 451–459.

Derstroff, M. C. (1995). *Mehrstufige Losgrößenplanung mit Kapazitätsbeschränkungen*. Physica,
Heidelberg.

Diaby, M., Bahl, H. C., Karwan, M. H., & Zionts, S. (1992). A Lagrangean Relaxation Approach
for Very-Large-Scale Capacitated Lot-Sizing. *Management Science*, *38*(9), 1329–1340.

Drechsel, J., & Kimms, A. (2009). An Algorithmic Approach to Cooperative Interval-Valued
Games and Interpretation Problems. *Business Research*, *2*(2), 206–213.

Drechsel, J., & Kimms, A. (2010a). Computing Core Cost Allocations for Cooperative Procurement. *International Journal of Production Economics, to appear*.

Drechsel, J., & Kimms, A. (2010b). Solutions and Fair Cost Allocations for Cooperative Lot Sizing
with Transshipments and Scarce Capacities. *International Journal of Production Research, to*
appear.

Drechsel, J., & Kimms, A. (2010c). The Subcoalition-Perfect Core of Cooperative Games. *Annals*
of Operations Research, to appear.

Drexl, A., & Kimms, A. (1997). Lot Sizing and Scheduling - Survey and Extensions. *European*
Journal of Operational Research, *99*(2), 221–235.

Dror, M., Guardiola, L. A., Meca, A., & Puerto, J. (2008). Dynamic Realization Games in
Newsvendor Inventory Centralization. *International Journal of Game Theory*, *37*(1), 139–153.

Dror, M., & Hartman, B. C. (2007). Shipment Consolidation: Who Pays for It and How Much?.
Management Science, *53*(1), 78–87.

Dubey, P. (1982). The Shapley Value as Aircraft Landing Fees-Revisited, *Management Science*.
28(8), 869–874.

Edgeworth, F. Y. (1881). *Mathematical Psychics – An Essay on the Application of Mathematics to*
the Moral Sciences. c. Keagon Paul & Co., London.

Engevall, S., Göthe-Lundgren, M., & Värbrand, P. (2004). The Heterogeneous Vehicle-Routing
Game. *Transportation Science*, *38*(1), 71–85.

Erlenkotter, D. (1978). A Dual-Based Procedure for Uncapacitated Facility Location. *Operations*
Research, *26*(6), 992–1009.

Essig, M. (2000). Purchasing Consortia as Symbiotic Relationships: Developing the Concept
of "Consortium Sourcing". *European Journal of Purchasing and Supply Management*, *6*(1),
13–22.

Estévez-Fernández, A., Borm, P., Meertens, M., & Reijnierse, H. (2009). On the Core of Routing
Games with Revenues. *International Journal of Game Theory*, *38*(2), 291–304.

Faigle, U. (1989). Cores of Games with Restricted Cooperation. *Mathematical Methods of Opera-*
tions Research, *33*(6), 405–422.

Faigle, U., Fekete, S. P., Hochstättler, W., & Kern, W. (1998a). On Approximately Fair Cost
Allocation in Euclidean TSP Games. *OR Spektrum*, *20*(1), 29–37.

Faigle, U., & Kern, W. (1992). The Shapley Value for Cooperative Games Under Precedence
Constraints. *International Journal of Game Theory*, *21*(3), 249–266.

Faigle, U., & Kern, W. (1993). On Some Approximately Balanced Combinatorial Cooperative Games. *Mathematical Methods of Operations Research, 38*(2), 141–152.

Faigle, U., & Kern, W. (1998). Approximate Core Allocation for Binpacking Games. *SIAM Journal on Discrete Mathematics, 11*(3), 387–399.

Faigle, U., & Kern, W. (2000). On the Core of Ordered Submodular Cost Games. *Mathematical Programming, 87*(3), 483–499.

Faigle, U., Kern, W., Fekete, S. P., & Hochstättler, W. (1997). On the Complexity of Testing Membership in the Core of Min-Cost-Spanning Tree Games. *International Journal of Game Theory, 26*(3), 361–366.

Faigle, U., Kern, W., Fekete, S. P., & Hochstättler, W. (1998b). The Nucleon of Cooperative Games and an Algorithm for Matching Games. *Mathematical Programming, 83*(1-3), 195–211.

Faigle, U., Kern, W., Kuipers, J. (2001). On the Computation of the Nucleolus of a Cooperative Game. *International Journal of Game Theory, 30*(1), 79–98.

Faigle, U., & Peis, B. (2006). *Cooperative Games with Lattice Structures*. Working Paper, Universität zu Köln.

Faigle, U., & Peis, B. (2008). A Hierarchical Model for Cooperative Games. In B. Monien, U.-P. Schroeder (Eds.), *SAGT 2008*, 230–241.

Fang, Q., Zhu, S., Cai, M., & Deng, X. (2002). On Computational Complexity of Membership Test in Flow Games and Linear Production Games. *International Journal of Game Theory, 31*(1), 39–45.

Federgruen, A., & Tzur, M. (1991). A Simple Forward Algorithm to Solve General Dynamic Lot Sizing Models with n Periods in O(n log n) or O(n) Time. *Management Science, 37*(8), 909–925.

Fisher, M. L. (1981). The Lagrangean Relaxation Method for Solving Integer Programming Problems. *Management Science, 27*(1), 1–18.

Florian, M., Bushell, G., Ferland, J., Guérin, G., & Nastansky, L. (1976). The Engine Scheduling Problem in a Railway Network. *INFOR, 14*(2), 121–138.

Frisk, M., Göthe-Lundgren, M., Jörnsten, K., & Rönnqvist, M. (2010). Cost Allocation in Collaborative Forest Transportation. *European Journal of Operational Research, 205*(2), 448–458.

Fromen, B. (2004). *Faire Aufteilung in Unternehmensnetzwerken*. Deutscher Universitäts-Verlag, Wiesbaden.

de Frutos, M. A. (1998). Decreasing Serial Cost Sharing under Economies of Scale. *Journal of Economic Theory, 79*(2), 245–275.

Fudenberg, D., & Tirole, J. (1991). *Game Theory*. MIT Presss, Cambridge, Massachusetts.

Gangolly, J. S. (1981). On Joint Cost Allocation: Independent Cost Proportional Scheme (ICPS) and Its Properties. *Journal of Accounting Research, 19*(2), 299–312.

Geoffrion, A. M. (1972). Generalized Benders Decomposition. *Journal of Optimization Theory and Applications, 10*(4), 237–260.

Geoffrion, A. M. (1974). Lagrangean Relaxation for Integer Programming. *Mathematical Programming Study, 2*(1), 81–114.

Geoffrion, A. M., & Graves, G. W. (1974). Multicommodity Distribution System Design by Benders Decomposition. *Management Science, 20*(5), 822–844.

Gerchak, Y., & Gupta, D. (1991). On Apportioning Costs to Customers in Centralized Continuous Review Inventory Systems. *Journal of Operations Management, 10*(4), 546–551.

Gilles, R. P., & Owen, G. (1999). *Cooperative Games and Disjunctive Permission Structures, centER Discussion Paper 9920*. Tilburg University.

Gilles, R. P., Owen, G., & van den Brink, R. (1992). Games with Permission Structures: The Conjunctive Approach. *International Journal of Game Theory, 20*(3), 277–293.

Gillies, D. B. (1959). Solutions to General Non-Zero-Sum Games. In A. W. Tucker, R. D. Luce (Eds.), *Contributions to the Theory of Games IV* pp. 47–85. Princeton University Press, Princeton.

Gilmore, P. C., & Gomory, R. E. (1961). A Linear Programming Approach to the Cutting-Stock Problem. *Operations Research, 9*(6), 849–859.

Goemans, M. X., & Skutella, M. (2004). Cooperative Facility Location Games, *Journal of Algorithms*, *50*(2), 194–214.

Gomory, R. E. (1958). Outline of an Algorithm for Integer Solutions to Linear Programs. *Bulletin of the American Mathematical Society*, *64*(5), 275–278.

González-Díaz, J., & Sánchez-Rodríguez, E. (2007). A Natural Selection from the Core of a TU Game: The Core-Center. *International Journal of Game Theory*, *36*(1), 27–46.

González-Díaz, J., & Sánchez-Rodríguez, E. (2009). Towards an Axiomatization of the Core-Center. *European Journal of Operational Research*, *195*(2), 449–459.

Göthe-Lundgren, M., Jörnsten, K., & Värbrand, P. (1996). On the Nucleolus of the Basic Vehicle Routing Game. *Mathematical Programming*, *72*(1), 83–100.

Grabisch, M., & Lange, F. (2007). Games on Lattices, Multichoice Games and the Shapley Value: A New Approach. *Mathematical Methods of Operations Research*, *65*(1), 153–167.

Grabisch, M., & Xie, L. (2007). A New Approach to the Core and Weber Set of Multichoice Games. *Mathematical Methods of Operations Research*, *66*(3), 491–512.

Grabisch, M., & Xie, L. (2008). *The Core of Games on Distributive Lattices: How to Share Benefits in Hierarchy*. Working Paper Centre d'Economie de la Sorbonne.

Grafe, F., Iñarra, E., & Zarzuelo, J. M. (1998). Population Monotonic Allocation Schemes on Externality Games. *Mathematical Methods of Operations Research*, *48*(1), 71–80.

Granot, D. (1986). A Generalized Linear Production Model: A Unifying Model. *Mathematical Programming*, *34*(2), 212–222.

Granot, D., & Granot, F. (1992). Computational Complexity of a Cost Allocation Approach to a Fixed Cost Spanning Forest Problem. *Mathematics of Operations Research*, *17*(4), 765–780.

Grünert, T. (1998). *Multi-Level Sequence-Dependent Dynamic Lot Sizing and Scheduling*. Shaker, Aachen.

Guardiola, L. A., Meca, A., & Puerto, J. (2006). *Coordination in Periodic Review Inventory Situations*. Working Paper Universidad Miguel Hernández de Elche.

Guardiola, L. A., Meca, A., & Puerto, J. (2008). Production-Inventory Games and PMAS Games: Characterizations of the Owen Point. *Mathematical Social Sciences*, *56*(1), 96–108.

Guardiola, L. A., Meca, A., & Puerto, J. (2009). Production-Inventory Games: A New Class of Totally Balanced Combinatorial Optimization Games. *Games and Economic Behavior*, *65*(1), 205–219.

Guardiola, L. A., Meca, A., & Timmer, J. (2007). Cooperation and Profit Allocation in Distribution Chains. *Decision Support Systems*, *44*(1), 17–27.

Hallefjord, A., Helming, R., & Jörnsten, K. (1995). Computing the Nucleolus when the Characteristic Function is Given Implicitly: A Constraint Generation Approach. *International Journal of Game Theory*, *24*(4), 357–372.

Hamers, H., Borm, P., van de Leensel, R., & Tijs, S. (1999). Cost Allocation in the Chinese Postman Problem. *European Journal of Operational Research*, *118*(1), 153–163.

Hamers, H., Klijn, F., Solymosi, T., Tijs, S., & Villar, J. P. (2002). Assignment Games Satisfy the CoMa-Property. *Games and Economic Behavior*, *38*(2), 231–239.

Hamiache, G. (1999). A Value with Incomplete Communication. *Games and Economic Behavior*, *26*(1), 59–78.

Hamlen, S. S., Hamlen, W. A., & Tschirhart, J. (1980). The Use of the Generalized Shapley Allocation in Joint Cost Allocation. *The Accounting Review*, *55*(2), 269–287.

Harrigan, K. R. (1988). Strategic Alliances and Partner Asymmetries. In F. J. Contractor, P. Lorange (Eds.), *Cooperative Strategies in International Business* pp. 205–226. Lexington Books, Lexington.

Harsanyi, J. C. (1966). A General Theory of Rational Behavior in Game Situations. *Econometrica*, *34*(3), 613–634.

Harsanyi, J. C. (1967). Games with Incomplete Information Played by Bayesian Players – Part I The Basic Model. *Management Science*, *14*(3), 159–182.

Harsanyi, J. C. (1968a). Games with Incomplete Information Played by Bayesian Players – Part II Bayesian Equilibrium Points. *Management Science*, *14*(5), 320–334.

Harsanyi, J. C. (1968b). Games with Incomplete Information Played by Bayesian Players – Part III The Basic Probability Distribution of the Game. *Management Science*, *14*(7), 486–502.

Harsanyi, J. C., & Selten, R. (1988). *A General Theory of Equilibrium Selection in Games*. MIT Press, Cambridge, Massachusetts.

Hartman, B. C., & Dror, M. (1996). Cost Allocation in Continuous-Review Inventory Models. *Naval Research Logistics*, *43*(4), 549–561.

Hartman, B. C., & Dror, M. (2003). Optimizing Centralized Inventory Operations in a Cooperative Game Theory Setting. *IIE Transactions*, *35*(3), 243–257.

Hartman, B. C., & Dror, M. (2005). Allocation of Gains from Inventory Centralization in Newsvendor Environments. *IIE Transactions*, *37*(2), 93–107.

Hartman, B. C., Dror, M., & Shaked, M. (2000). Cores of Inventory Centralization Games. *Games and Economic Behavior*, *31*(1), 26–49.

Heijboer, G. (2002). Allocating Savings in Purchasing Consortia: Analysing Solutions from a Game Theoretic Perspective. In *Proceedings of the 11th International Annual IPSERA Conference, Enschede* 275–287.

Heijboer, G. (2003). *Quantitative Analysis of Strategic and Tactical Purchasing Decisions*. Twente University Press, Enschede.

Helber, S., & Sahling, F. (2010). A Fix-and-Optimize Approach for the Multi-Level Capacitated Lot Sizing Problem. *International Journal of Production Economics*, *123*(2), 247–256.

Held, M., Wolfe, P., & Crowder, H. P. (1974). Validation of Subgradient Optimization. *Mathematical Programming*, *6*(1), 62–88.

Hendrick, T. E. (1997). Purchasing Consortiums: Horizontal Alliances among Firms Buying Common Goods and Services – What? Who? Why? How?, cAPS Focus Study. www.capsresearch.org.

van den Heuvel, W., Borm, P., & Hamers, H. (2007a). Economic Lot-Sizing Games. *European Journal of Operational Research*, *176*(2), 1117–1130.

van den Heuvel, W., Kundakcioglu, O. E., Geunes, J., Romeijn, H. E., Sharkey, T. C., & Wagelmans, A. P. M. (2007b). *Integrated Market Selection and Production Planning Complexity and Solution Approaches*. Working Paper Erasmus University Rotterdam.

Hinojosa, M. A., Mármol, A. M., & Thomas, L. C. (2005). Core, Least Core, and Nucleolus for Multiple Scenario Cooperative Games. *European Journal of Operational Research*, *164*(1), 225–238.

Hsiao, C.-R. (1995a). A Value for Continuously-Many-Choice Cooperative Games. *International Journal of Game Theory*, *24*(3), 273–292.

Hsiao, C.-R.(1995b). A Note on Non-Essential Players in Multichoice Cooperative Games. *Games and Economic Behavior*, *8*(2), 424–432.

Hsiao, C.-R., & Raghavan, T. E. S. (1992). Monotonicity and Dummy Free Property for Multichoice Cooperative Games. *International Journal of Game Theory*, *21*(3), 301–312.

Hsiao, C.-R., & Raghavan, T. E. S. (1993). Shapley Value for Multichoice Cooperative Games. *Games and Economic Behavior*, *5*(2), 240–256.

Hwang, Y.-A., & Liao, Y.-H. (2008). Potential in Multichoice Cooperative TU Games. *Asia-Pacific Journal of Operational Research*, *25*(5), 591–611.

Ichiishi, T. (1981). Super-Modularity: Applications to Convex Games and to the Greedy Algorithm for LP. *Journal of Economic Theory*, *25*(2), 283–286.

Illner, A. (1999). *Kostenallokation bei dezentraler Organisation*. Peter Lang Europäischer Verlag der Wissenschaften, Frankfurt.

Jain, K., & Mahdian, M. (2007). Cost Sharing. In N. Nisan, T. Roughgarden, V. Tardos, V. V. Vazirani (Eds.), *Algorithmic Game Theory* pp. 385–410. Cambridge University Press, Cambridge.

Johnson, F. P. (1999). The Pattern of Evolution in Public Sector Purchasing Consortia. *International Journal of Logistics: Research and Applications*, *2*(1), 57–73.

Jost, P.-J. (Ed.). (2001). *Die Spieltheorie in der Betriebswirtschaftslehre*. Schäffer-Poeschel Verlag, Stuttgart.

Kamiya, K., & Talman, D. (1991). Simplicial Algorithm for Computing a Core Element in a Balanced Game. *Journal of the Operations Research Society of Japan, 34*(2), 222–228.

Kaneko, M., & Wooders, M. H. (1982). Cores of Partitioning Games. *Mathematical Social Sciences, 3*(4), 313–327.

Kannai, Y. (1999). The Core and Balancedness, In R. J. Aumann, S. Hart (Eds.), *Handbook of Game Theory with Economic Applications, Volume 1* pp. 355–395. Elsevier Science Publishers B. V., Amsterdam.

Karimi, B., Fatemi Ghomi, S. M. T., & Wilson, J. M. (2003). The Capacitated Lot Sizing Problem: A Review of Models and Algorithms. *Omega, 31*(5), 365–378.

Klijn, F., & Slikker, M. (2005). Distribution Center Consolidation Games. *Operations Research Letters, 33*(3), 285–288.

Klijn, F., Slikker, M., & Zarzuelo, J. (1999). Characterizations of a Multichoice Value. *International Journal of Game Theory, 28*(4), 521–532.

Kogan, K., & Tapiero, C. S. (2007). *Supply Chain Games: Operations Management and Risk Valuation*. Springer, New York.

de Kok, A. G., & Graves, S. C. (2003). Introduction. In A. G. de Kok, S. C. Graves (Eds.), *Supply Chain Management: Design, Coordination and Operation* pp. 1–14. Elsevier, Amsterdam.

Krajewska, M. A., Kopfer, H., Laporte, G., Ropke, S., & Zaccour, G. (2008). Horizontal Cooperation among Freight Carriers: Request Allocation and Profit Sharing. *Journal of the Operational Research Society, 59*(11), 1483–1491.

Kuipers, J. (1993). On the Core of Information Graph Games. *International Journal of Game Theory, 21*(4), 339–350.

Lang, J. C., & Domschke, W. (2010). Efficient Reformulations for Dynamic Lot-Sizing Problems with Product Substitution. *OR Spectrum, 32*(2), 263–291.

Leach, W. D. (2006). Collaborative Public Management and Democracy: Evidence from Western Watershed Partnerships. *Public Administration Review, 66*(1), 100–110.

Lehrer, E. (2002). Allocation Processes in Cooperative Games. *International Journal of Game Theory, 31*(3), 341–351.

Lemaire, J. (1984). An Application of Game Theory: Cost Allocation. *Astin Bulletin, 14*(1), 61–81.

Leng, M., & Parlar, M. (2005). Game Theoretic Applications in Supply Chain Management: A Review. *INFOR, 43*(3), 187–220.

Lewis, J. D. (1990). *Partnerships for Profit – Structuring and Managing Strategic Alliances*. The Free Press, New York.

Lewontin, R. C. (1961). Evolution and the Theory of Games. *Journal of Theoretical Biology, 1*(3), 382–403.

Littlechild, S. C. (1974). A Simple Expression for the Nucleolus in a Special Case. *International Journal of Game Theory, 3*(1), 21–29.

Littlechild, S. C., & Thompson, G. F. (1973). A Simple Expression for the Shapley Value in a Special Case. *Management Science, 20*(3), 370–372.

Littlechild, S. C., & Thompson, G. F. (1977a). Aircraft Landing Fees: A Game Theory Approach. *The Bell Journal of Economics, 8*(2), 186–204.

Littlechild, S. C., & Thompson, G. F. (1977b). A Further Note on the Nucleolus of the "Airport Game". *International Journal of Game Theory, 5*(2/3), 91–95.

Littlechild, S. C., & Vaidya, K. G. (1976). The Propensity to Disrupt and the Disruption Nucleolus of a Charachteristic Function Game. *International Journal of Game Theory, 5*(2/3), 151–161.

Louderback, J. G. (1976). Another Approach to Allocating Joint Costs: A Comment. *The Accounting Review, 51*(3), 683–685.

Love, S. F. (1973). Bounded Production and Inventory Models with Piecewise Concave Costs. *Management Science, 20*(3), 313–318.

Luce, R. D., & Raiffa, H. (1957). *Games and Decisions – Introduction and Critical Survey*. John Wiley & Sons, Inc., New York.

Magnanti, T. L., & Wong, R. T. (1981). Accelerating Benders Decomposition: Algorithmic Enhancement and Model Selection Criteria. *Operations Research, 29*(3), 464–484.

Maschler, M. (1992). The Bargaining Set, Kernel, and Nucleolus. In R. J. Aumann, S. Hart (Eds.), *Handbook of Game Theory with Economic Applications, Volume 1* pp. 591–667. Elsevier Science Publishers B. V., Amsterdam.

Maschler, M., Peleg, B., & Shapley, L. S. (1979). Geometric Properties of the Kernel, Nucleolus, and Related Solution Concepts. *Mathematics of Operations Research, 4*(4), 303–338.

Meca, A. (2007). A Core-Allocation Family for Generalized Holding Cost Games. *Mathematical Methods of Operations Research, 65*(3), 499–517.

Meca, A., García-Jurado, I., & Borm, P. (2003). Cooperation and Competition in Inventory Games. *Mathematical Methods of Operations Research, 57*(3), 481–493.

Meca, A., Guardiola, L. A., & Toledo, A. (2007). *p*-Additive Games: A Class of Totally Balanced Games Arising from Inventory Situations with Temporary Discounts. *TOP, 15*(2), 322–340.

Meca, A., & Timmer, J. (2008). Supply Chain Collaboration. In V. Kordic (Ed.), *Supply Chain Theory and Applications* pp. 1–18. I-Tech Education and Publishing, Vienna.

Meca, A., Timmer, J., García-Jurado, I., & Borm, P. (2004). Inventory Games. *European Journal of Operational Research, 156*(1), 127–139.

Millar, H. H., & Yang, M. (1994). Lagrangean Heuristics for the Capacitated Multi-Item Lot-Sizing Problem with Backordering. *International Journal of Production Economics, 34*(1), 1–15.

Minner, S. (2007). Bargaining for Cooperative Economic Ordering. *Decision Support Systems, 43*(2), 569–583.

Moghaddam, A. T. N., & Michelot, C. (2009). A Contribution to the Linear Programming Approach to Joint Cost Allocation: Methodology and Application. *European Journal of Operational Research, 197*(3), 999–1011.

Monczka, R. M., Trent, R. J., & Handfield, R. B. (2002). *Purchasing and Supply Chain Management* (2nd ed.). South-Western Thomson Learning.

Montrucchio, L., & Scarsini, M. (2007). Large Newsvendor Games. *Games and Economic Behavior, 58*(2), 316–337.

Morgenstern, O. (1963). *Spieltheorie und Wirtschaftswissenschaft*. Oldenbourg, Wien.

Moriarity, S. (1975). Another Approach to Allocating Joint Costs. *The Accounting Review, 50*(4), 791–795.

Moriarity, S. (1981a). Some Rationales for Cost Allocations. In S. Moriarity (Ed.), *Joint Cost Allocations – Proceedings of the University of Oklahoma, Conference on Cost Allocation* 8–13.

Moriarity S. (Ed.). (1981b). *Joint Cost Allocations*. Proceedings of the University of Oklahoma, Conference on Cost Allocation, Oklahoma.

Mosquera, M. A., García-Jurado, I., & Fiestras-Janeiro, M. G. (2008). A Note on Coalitional Manipulation and Centralized Inventory Management. *Annals of Operations Research, 158*(1), 183–188.

Moulin, H., & Shenker, S. (1992). Serial Cost Sharing. *Econometrica, 60*(5), 1009–1037.

Müller, A., Scarsini, M., & Shaked, M. (2002). The Newsvendor Game Has a Nonempty Core. *Games and Economic Behavior, 38*(1), 118–126.

Muto, S., Nakayama, M., Potters, J., & Tijs, S. (1988). On Big Boss Games. *The Economics Studies Quarterly, 39*(4), 303–321.

Muto, S., Potters, J., & Tijs, S. (1989). Information Market Games. *International Journal of Game Theory, 18*(2), 209–226.

Myerson, R. B. (1977). Graphs and Cooperation in Games. *Mathematics of Operations Research, 2*(2), 225–229.

Myerson, R. B. (1991). *Game Theory: Analysis of Conflict*. Harvard University Press, Cambridge Massachusetts.

Nagarajan, M., & Sošić, G. (2008). Game-Theoretic Analysis of Cooperation among Supply Chain Agents: Review and Extensions. *European Journal of Operational Research, 187*(3), 719–745.

Nagarajan, M., Sošić, G., & Zhang, H. (2008). Stability of Group Purchasing Organizations. Working Paper.

Nash, J. F. (1950). Equilibrium Points in *n*-Person Games. In *Proceedings of the National Academy of Sciences of the United States of America (36)* 48–49.

Nash, J. F. (1951). Non-Cooperative Games. *Annals of Mathematics, 54*(2), 286–295.

Nash, J. F. (1953). Two-Person Cooperative Games. *Econometrica, 21*(1), 128–140.

von Neumann, J. (1928). Zur Theorie der Gesellschaftsspiele. *Mathematische Annalen, 100*(1), 295–320.

von Neumann, J., & Morgenstern, O. (2004). *Theory of Games and Economic Behavior* (Sixtieth-Anniversary ed.). Princeton University Press, Princeton.

Nisan, N., Roughgarden, T., Tardos, V., & Vazirani, V. V. (Eds.). (2007). *Algorithmic Game Theory*. Cambridge University Press, Cambridge.

van den Nouweland, A., Tijs, S., Potters, J., & Zarzuelo, J. (1995). Cores and Related Solution Concepts for Multi-Choice Games. *Mathematical Methods of Operations Research, 41*(3), 289–311.

Núñez, M., & Rafels, C. (1998). On Extreme Points of the Core and Reduced Games. *Annals of Operations Research, 84*(1), 121–133.

Okamoto, Y. (2003). Some Properties of the Core on Convex Geometries. *Mathematical Methods of Operations Research, 56*(3), 377–386.

Osborne, M. J., & Rubinstein, A. (1994). *A Course in Game Theory*. MIT Press, Cambridge Massachusetts.

Owen, G. (1975). On the Core of Linear Production Games. *Mathematical Programming, 9*(1), 358–370.

Owen, G. (1986). Values of Graph-Restricted Games. *SIAM Journal of Algebraic and Discrete Methods, 7*(2), 210–220.

Owen, G. (2001). *Game Theory* (3rd ed.). Academic Presss, San Diego.

Özdamar, L., & Barbarosoğlu (1999). Hybrid Heuristics for the Multi-Stage Capacitated Lot Sizing and Loading Problem. *Journal of the Operational Research Society, 50*(8), 810–825.

Özen, U., Erkip, N., & Slikker, M. (2006). *Profit Division in Newsvendor Situations with Delivery Restrictions*. Working Paper Eindhoven University of Technology.

Özen, U., Fransoo, J., Norde, H., & Slikker, M. (2008). Cooperation between Multiple Newsvendors with Warehouses. *Manufacturing & Service Operations Management, 10*(2), 311–324.

Özen, U., Norde, H., & Slikker, M. (2010). On the Convexity of Newsvendor Games. *International Journal of Production Economics*, online appearance February 2010.

Özen, U., & Sošic, G. (2006). *A Multi-Retailer Decentralized Distribution System with Updated Demand Information*. Working Paper Eindhoven University of Technology.

Özen, U., Sošic, G., & Slikker, M. (2007). A Collaborative Decentralized Distribution System with Demand Updates. Working Paper.

Özener, O. O., & Ergun, O. (2008). Allocating Costs in a Collaborative Transportation Procurement Network. *Transportation Science, 42*(2), 146–165.

Papadimitriou, C. H. (2007). The Complexity of Finding Nash Equilibria. In N. Nisan, T. Roughgarden, V. Tardos, V. V. Vazirani (Eds.), *Algorithmic Game Theory*. Cambridge University Press, Cambridge, 29–51.

Peleg, B. (1999). Axiomatizations of the Core. In R. J. Aumann, S. Hart (Eds.), *Handbook of Game Theory with Economic Applications, Volume 1* pp. 397–412. Elsevier Science Publishers B. V., Amsterdam.

Peleg, B., & Sudhölter, P. (2007). *Introduction to the Theory of Cooperative Games* (2nd ed.). Springer, Berlin.

Peters, H., & Zank, H. (2005). The Egalitarian Solution for Multichoice Games. *Annals of Operations Research, 137*(1), 399–409.

Pfaff, D. (1994). On the Allocation of Overhead Costs. *European Accounting Review, 1*(1), 49–70.

Pochet, Y., & Wolsey, L. A. (2006). *Production Planning by Mixed Integer Programming*. Springer, Heidelberg.

Porter, M. E. (1985). *Competitive Advantage – Creating and Sustaining Superior Performance*. The Free Press, New York.

Potters, J., & Sudhölter, P. (1999). Airport Problems and Consistent Allocation Rules. *Mathematical Social Sciences, 38*(1), 83–102.

Pulido, M. A., & Sánchez-Soriano, J. (2006). Characterization of the Core in Games with Restricted Cooperation. *European Journal of Operational Research, 175*(2), 860–869.

Pyke, D. F., & Johnson, M. E. (2004). Sourcing Strategy and Supplier Relationships: Alliances versus eProcurement. In T. P. Harrison, H. L. Lee, J. J. Neale (Eds.), *The Practice of Supply Chain Management: Where Theory and Application Converge* pp. 77–89. Springer, New York.

Quadt, D., & Kuhn, H. (2008). Capacitated Lot-Sizing with Extensions: A Review. *4OR*, *6*(1), 61–83.

Quindt, T. (1991). Necessary and Sufficient Conditions for Balancedness in Partitioning Games. *Mathematical Social Sciences*, *22*(1), 87–91.

Ransmeier, J. S. (1942). *The Tennessee Valley Authority*. The Vanderbilt University Press, Nashville.

Rasmusen, E. (2007). *Games and Information – An Introduction to Game Theory* (4th ed.). Blackwell Publishing, Malden.

Roth, A. E. (Ed.). (1988). *The Shapley Value – Essays in Honor of Lloyd S. Shapley*. Cambridge University Press, Cambridge.

Sahling, F., Buschkühl, L., Tempelmeier, H., & Helber, S. (2009). Solving a Multi-Level Capacitated Lot Sizing Problem with Multi-Period Setup Carry-Over via a Fix-and-Optimize Heuristic. *Computers & Operations Research*, *36*(9), 2546–2553.

Sambasivan, M., & Yahya, S. (2005). A Lagrangean-Based Heuristic for Multi-Plant, Multi-Item, Multi-Period Capacitated Lot-Sizing Problems with Inter-Plant Transfers. *Computers & Operations Research*, *32*(3), 537–555.

Sánchez-Soriano, J. (2006). Pairwise Solutions and the Core of Transportation Situations. *European Journal of Operational Research*, *175*(1), 101–110.

Schmeidler, D. (1969). The Nucleolus of a Characteristic Function Game. *SIAM Journal of Applied Mathematics*, *17*(6), 1163–1170.

Schotanus, F. (2007). *Horizontal Cooperative Purchasing*. PrintPartners Ipskamp B. V., Enschede.

Schotanus, F., Telgen, J., & de Boer, L. (2008). Unfair Allocation of Gains under the Equal Price Allocation Method in Purchasing Groups. *European Journal of Operational Research*, *187*(1), 162–176.

Schrijver, A. (1986). *Theory of Linear and Integer Programming*. John Wiley & Sons, Chichester.

Schulz, A. S., & Uhan, N. A. (2007). Encouraging Cooperation in Sharing Supermodular Costs. In *Approximation, Randomization, and Combinatorial Optimization. Algorithms and Techniques*, 271–285.

Selten, R. (1965). Spieltheoretische Behandlung eines Oligopolmodells mit Nachfrageträgheit. *Zeitschrift für die gesamte Staatswissenschaft*, *121*(2 and 4), 301–324 and 667–689.

Selten, R. (1975). Reexamination of the Perfectness Concept for Equilibrium Points in Extensive Games. *International Journal of Game Theory*, *4*(1), 25–55.

Sethi, A. P., & Thompson, G. L. (1984). The Pivot and Probe Algorithm for Solving a Linear Program. *Mathematical Programming*, *29*(2), 219–233.

Shapley, L. S. (1953). A Value for *n*-Person Games. In: A. W. Tucker, H. W. Kuhn (Eds.), *Contributions to the Theory of Games II* pp. 307–317. Princeton University Press, Princeton.

Shapley, L. S. (1967). On Balanced Sets and Cores. *Naval Research Logistics Quarterly*, *14*(4), 453–460.

Shapley, L. S. (1971). Cores of Convex Games. *International Journal of Game Theory*, *1*(1), 11–26.

Shapley, L. S., & Shubik, M. (1966). Quasi-Cores in a Monetary Economy with Nonconvex Preferences. *Econometrica*, *34*(4), 805–827.

Shapley, L. S., & Shubik, M. (1971). The Assignment Game I: The Core. *International Journal of Game Theory*, *1*(1), 111–130.

Shenkar, O., & Reuer, J. J. (Eds.). (2006). *Handbook of Strategic Alliances*. Sage Publications, Thousand Oaks.

Shubik, M. (1962). Incentives, Decentralized Control, the Assignment of Joint Costs and Internal Pricing. *Management Science*, *8*(3), 325–343.

Shubik, M. (1982). *Game Theory in the Social Sciences*. MIT Press, Cambridge, Massachusetts.

Simatupang, T. M., & Sridharan, R. (2002). The Collaborative Supply Chain. *International Journal of Logistics Management*, *13*(1), 15–30.

Simatupang, T. M., & Sridharan, R. (2005). The Collaboration Index: A Measure for Supply Chain Collaboration. *International Journal of Physical Distribution & Logistics Management, 35*(1), 44–62.

Simchi-Levi, D., Kaminsky, P., & Simchi-Levi, E. (2004). *Managing the Supply Chain*. McGraw-Hill, New York.

Slikker, M., Fransoo, J., & Wouters, M. (2001). Joint Ordering in Multiple News-Vendor Problems: A Game-Theoretical Approach. Working Paper Eindhoven University of Technology.

Slikker, M., Fransoo, J., & Wouters, M. (2005). Cooperation between Multiple News-Vendors with Transshipments. *European Journal of Operational Research, 167*(2), 370–380.

Snyder, H., & Davenport, E. (1997). What Does It Really Cost? Allocating Indirect Costs. *Asian Libraries, 6*(3/4), 205–214.

Solymosi, T. (1999). On the Bargaining Set, Kernel and Core of Superadditive Games. *International Journal of Game Theory, 28*(2), 229–240.

Solymosi, T. (2008). Bargaining Sets and the Core in Partitioning Games. *Central European Journal of Operations Research, 16*(4), 425–440.

Sotomayor, M. (2003). Some further Remark on the Core Structure of the Assignment Games. *Mathematical Social Sciences, 46*(3), 261–265.

Soumis, F. (1997). Decomposition and Column Generation. In M. Dell'Amico, F. Maffioli, S. Martello (Eds.), *Annotated Bibliographies in Combinatorial Optimization*. John Wiley & Sons, New York, 115–126.

Sox, C. R., & Gao, Y. (1999). The Capacitated Lot Sizing Problem with Setup Carry-Over. *IIE Transactions, 31*(2), 173–181.

Stadtler, H. (2003). Multilevel Lot Sizing with Setup Times and Multiple Constrained Resources: Internally Rolling Schedules with Lot-Sizing Windows. *Operations Research, 51*(3), 487–502.

Stadtler, H. (2009). A Framework for Collaborative Planning and State-of-the-Art. *OR Spectrum, 31*(1), 5–30.

Straffin, P. D., & Heaney, J. P. (1981). Game Theory and the Tennessee Valley Authority. *International Journal of Game Theory, 10*(1), 35–43.

Tamir, A. (1992). On the Core of Cost Allocation Games Defined on Location Problems. *Transportation Science, 27*(1), 81–86.

Tempelmeier, H., & Derstroff, M. (1996). A Lagrangean-Based Heuristic for Dynamic Multilevel Multiitem Constrained Lotsizing with Setup Times. *Management Science, 42*(5), 738–757.

Thompson, G. L., & Sethi, A. P. (1986). Solution of Constrained Generalized Transportation Problems Using the Pivot and Probe Algorithm. *Computers & Operations Research, 13*(1), 1–9.

Thun, J.-H. (2005). The Potential of Cooperative Game Theory for Supply Chain Management. In H. Kotzab, S. Seuring, M. Müller, G. Reiner, (Eds.), *Research Methodologies in Supply Chain Management* pp. 477–491. Physica-Verlag, Heidelberg.

Tijs, S. H., & Driessen, S. H. (1986). Extensions of Solution Concepts by Means of Multiplicative ϵ-Tax Games. *Mathematical Social Sciences, 12*(1), 9–20.

Todeva, E., & Knoke, D. (2005). Strategic Alliances and Models of Collaboration. *Management Decision, 43*(1), 123–148.

van Velzen, B., Hamers, H., & Norde, H. (2004). Characterizing Convexity of Games Using Marginal Vectors. *Discrete Applied Mathematics, 143*(3), 298–306.

Völker, R., & Neu, J. (2008). *Supply Chain Collaboration*. Physica-Verlag, Heidelberg.

Wagelmans, A., van Hoesel, S., & Kolen, A. (1992). Economic Lot Sizing: An O(n log n) Algorithm that Runs in Linear Time in the Wagner-Whitin Case. *Operations Research, 40*(1), S145–S156.

Wagner, H. M., & Whitin, T. M. (1958). Dynamic Version of the Economic Lot Size Model. *Management Science, 5*(1), 89–96.

Weber, R. J. (1978). Probabilistic Values for Games. Discussion Paper Yale University.

Xie, L. (2006). Some Solutions of Regular Capacities and Regular Games. In *Proceedings International Conference on Information Processing and Management of Uncertainty (IPMU 06)* 1768–1773.

Xu, D., & Yang, R. (2009). A Cost-Sharing Method for an Economic Lot-Sizing Game. *Operations Research Letters*, *37*(2), 107–110.

Young, H. P. (1985). Monotonic Solutions of Cooperative Games. *International Journal of Game Theory*, *14*(2), 65–72.

Young, H. P. (1994). Cost Allocation. In: R. J. Aumann, S. Hart (Eds.), *Handbook of Game Theory with Economic Applications, Volume 2* pp. 1193–1235. Elsevier Science Publishers B. V., Amsterdam.

Young, H. P., Okada, N., & Kashimoto, T. (1982). Cost Allocation in Water Resources Development. *Water Resources Research*, *18*(3), 463–475.

Zangwill, W. I. (1969). A Backlogging Model and a Multi-Echelon Model of a Dynamic Economic Lot Size Production System – A Network Approach. *Management Science*, *18*(9), 506–527.

Zelewski, S. (2007). Faire Verteilung von Effizienzgewinnen in Supply Webs. In H. Corsten, H. Missbauer (Eds.), *Produktions- und Logistikmanagement – Festschrift für Günther Zäpfel zum 65. Geburtstag* pp. 553–572. Verlag Franz Vahlen, München.

Zermelo, E. (1913). Über eine Anwendung der Mengenlehre auf die Theorie des Schachspiels. In E. W. Hobson, A. E. H. Love (Eds.), *Proceedings of the Fith International Congress of Mathematicians*. Cambridge University Press, Cambridge, 501–504.

Zhang, J. (2009). Cost Allocation for Joint Replenishment Models. *Operations Research*, *57*(1), 146–156.